Electronic Circuits Cookbook

Electronic Circuits Cookbook

Harry L. Helms

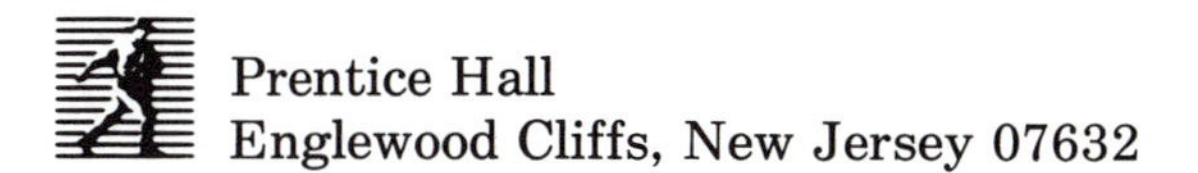

Prentice Hall
Englewood Cliffs, New Jersey 07632

Library of Congress
Library of Congress Cataloging-in-Publication Data

Helms, Harry L.
Electronic circuits cookbook / Harry L. Helms.
p. cm.

ISBN 0-13-250168-6
1. Electronics—Amateurs' manuals. I. Title.
TK9965.H356 1988
621.3815'3—dc19 87-34047
CIP

Editorial/production supervision and
interior design: **WordCrafters Editorial Services, Inc.**
Cover design: **Diane Saxe**
Manufacturing buyer: **P. Benevento**

This book can be made available to businesses and organizations at a special discount when ordered in large quantities. For more information contact:

Prentice Hall Inc.
Special Sales and Markets
College Division
Englewood Cliffs, NJ 07632

Printed in the United States of America

10 9 8 7 6 5 4 3 2 1

ISBN 0-13-250168-6

Prentice-Hall International (UK) Limited, *London*
Prentice-Hall of Australia Pty. Limited, *Sydney*
Prentice-Hall Canada Inc., *Toronto*
Prentice-Hall Hispanoamericana, S.A., *Mexico*
Prentice-Hall of India Private Limited, *New Delhi*
Prentice-Hall of Japan, Inc., *Tokyo*
Simon & Schuster Asia Pte. Ltd., *Singapore*
Editora Prentice-Hall do Brasil, Ltda., *Rio de Janeiro*

This book is dedicated to Jonathan Erickson, Stan Miastkowski, and other fellow members of RSTPAA (Radio Shack Technical Publications Alumni Association) worldwide

Contents

Introduction

Why reinvent the wheel? Why repeat the effort someone else has expended?

If you do much work at all with electronics, you've probably collected several circuits for various purposes. These are circuits you frequently need, or circuits which can serve as a beginning point for your own designs. Or they may be circuits that are just "interesting," or ones you plan to get around to building and working with someday.

But collecting and using these circuits can be a problem. It would be nice to have the circuits collected together in one place, with supporting data, organized and indexed so the desired circuit can be readily found. Stuffing them into a file folder or manila envelope leaves something to be desired.

That's the philosophy behind *Electronic Circuits Cookbook.* This is a collection of "prefab" circuits that are ready to go. Many can be built to accomplish certain functions by altering the values of certain components, and full design equations have been included where possible. However, you won't find extensive theory or explanations about circuit operation, although some basic background is included where appropriate. After all, this is a *cookbook:* add the components indicated together as shown in the circuit diagram, and you get a specified result. If you want to know about the "chemistry" involved, there's no shortage of excellent texts and references you can consult. Make no mistake about it—this book emphasizes the "how," but not the "why." But it does give you all you need to know to get a circuit up and running—and use it—in the shortest possible time.

The emphasis throughout the book is on the practical and efficient (okay, the "easy"). Thus, integrated devices are used in place of discrete semiconductors whenever possible, and common ICs are used in place of more obscure devices unless the obscure device offers substantially better performance or is the only device available for a given purpose. Also, pin numbers are given for ICs, and specific supply voltages are indicated for some circuits, while in other cases "V," "V_{cc}," or "V_{dd}" is specified. In the latter the supply voltage can be anywhere in the range allowed for the IC used, i.e., +5 V for TTL, +3 to +15 V for CMOS, and +9 V (in single or dual polarity, as indicated) for most linear devices. Efforts have been made to avoid circuits requiring oddball or precision components (such as 1 percent tolerance resistors), but in some cases the use of such parts is unavoidable. In other circuits, nonstandard values of resistance and capacitance are indicated; these values can be created by combining standard parts values in series or parallel.

The emphasis on the practical does not mean that all circuit configurations in this book are deadly serious. Indeed, some have been included strictly because they are fun to build and use. (After all, what's the point of being involved with electronics if you can't have a little fun with it?) Student readers of this book will find several of the circuits useful for class projects and science fairs.

Of course, normal handling and operating conditions for the various ICs must be observed. In particular, this means that CMOS devices should be handled so as to avoid damage due to static discharge and TTL devices should be used with decoupling capacitors. Also, all unused inputs on logic devices should be tied to either a high or a low logic level to prevent erratic operation of the device, and the input voltages to a device usually must not exceed the supply voltage.

Since the circuits included in this book have been selected for their generality, many of them will not be the optimal solution to a specific application or problem. However, they can be used as the starting design for a more appropriate circuit or as a "patch" until a more suitable circuit is developed. Although the circuits were selected with their adaptability and potential for modification in mind, this goal was not always achieved. Nonetheless, in most cases it was realized to an adequate degree.

Several of the circuits, such as transmitters and oscillators, produce RF energy and thus are subject to the rules and regulations of the Federal Communications Commission (or similar agencies in other countries). Some of the circuits can be used in the United States without an FCC license, while others require an appropriate amateur or commercial operator's license before an antenna can be connected. (Operation into a resistive "dummy load" is okay, however.) Any restrictions which apply to the operation of a transmitter or oscillator circuit will be noted in the text accompanying that circuit.

The circuits shown are ideal for "playing around" and experimentation, since many can be built on a conventional solderless breadboard. Of course, common good breadboarding practice should be followed. In particular, long wires and component leads in circuits operating at higher frequencies should be avoided. Placement of components can become crucial in RF and high-speed digital circuits; certain circuits, such as those involving higher frequency RF energy, simply cannot be built using solderless breadboard techniques. Instead, "perfboard," conventional printed circuit boards, or even special "RF breadboards" must be used in such cases.

I hope that you'll find this book a useful addition to your electronics library and that it saves you hours of time and effort in locating circuits you may need.

Harry L. Helms

a

Active filters

One of the first applications of operational amplifiers was in active filters, where the gain they afforded allowed markedly superior results than those achieved by filters employing strictly passive components. There are three main active filter configurations. The *high-pass* filter allows all frequencies above a certain point, known as the *cutoff frequency,* to pass without attenuation while suppressing all frequencies below the cutoff. The *low-pass* filter does the opposite: all frequencies below the cutoff can pass, but all above are rejected. Finally, the *bandpass* filter permits a range of frequencies centered on a *center frequency* to pass while rejecting all frequencies above and below the permitted range.

Figure 1 shows a diagram of a high-pass filter whose cutoff frequency depends upon the values of capacitors $C1$ and $C2$ as well as resistors $R1$, $R2$, $R3$, and $R4$. $C1$ and $C2$ are equal in value, as are $R1$ and $R2$. The formula for determining the cutoff frequency is

$$\text{cutoff frequency} = \frac{1}{2\pi RC}$$

where $\pi = 3.14159$, R is equal to either $R1$ or $R2$, and C is equal to either $C1$ or $C2$. The gain of the filter is equal to $R4$ divided by $R3$ and the output at the cutoff frequency is equal to 0.707 that of the circuit's maximum output.

A few minor changes to the circuit of Figure 1 produce a low-pass filter, the schematic of which is shown in Figure 2. The same formulas used with high-pass filters for determining the cutoff frequency and the gain at cutoff apply to low-pass filters.

Figure 3 shows the schematic for a bandpass filter. The center frequency of the circuit is determined by the setting of the 1 K potentiometer. The greater the resistance selected, the lower the cen-

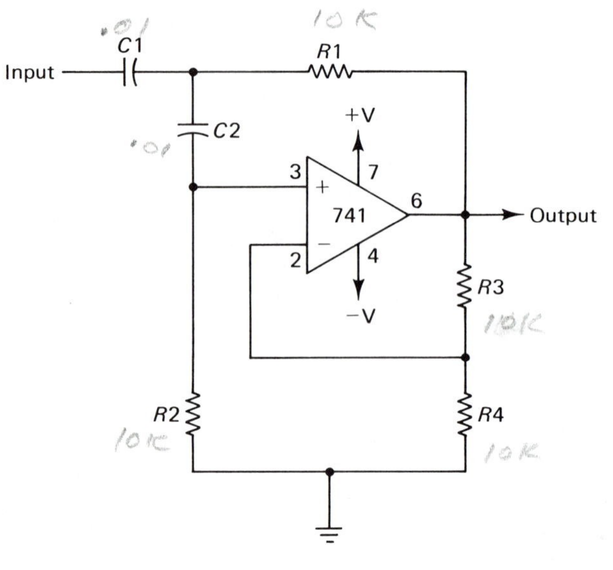

Figure 1 High-pass filter

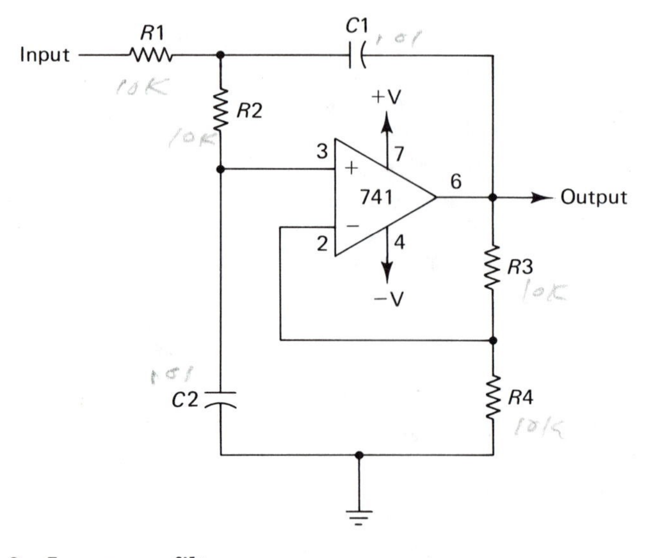

Figure 2 Low-pass filter

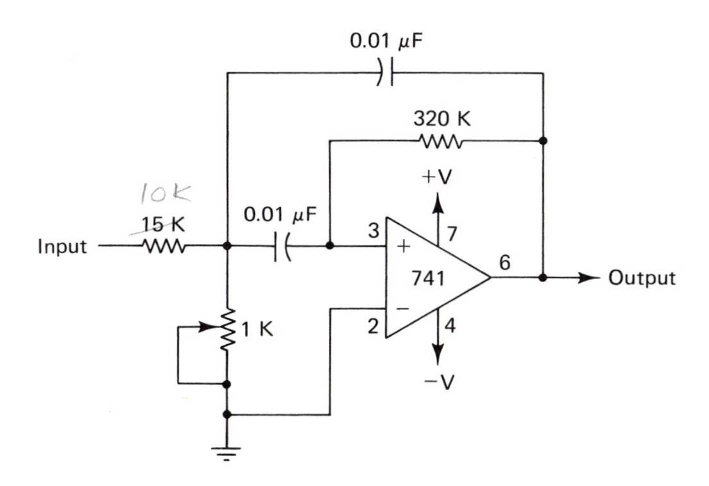

Figure 3 Bandpass filter

ter frequency. For example, a setting of approximately 930 Ω produces a center frequency of 1,000 Hz, while a setting of about 130 Ω gives a center frequency of 2,000 Hz. A fixed resistor can be substituted if only one center frequency is needed, or a potentiometer of greater value may be used if center frequencies below 1,000 Hz are required.

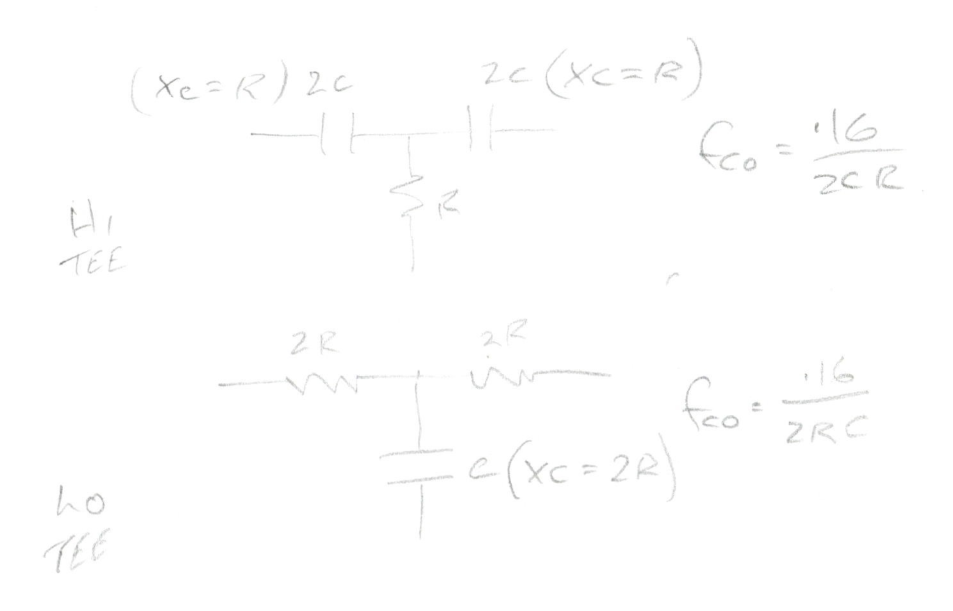

Alarm, high/low limit

The circuit shown in Figure 4 will produce a square wave output depending upon the input voltage, an upper voltage limit, and a lower voltage limit. The values of the upper and lower voltage limits depend upon the value of certain components used in the circuit. The circuit will oscillate and produce an output if the input voltage is greater than the upper limit or if the input voltage is less than the lower limit; otherwise, the output will be a low logic level. The frequency of the square wave output produced when the input voltage is beyond the lower and upper limits also depends upon the values of certain of the circuit's components.

The values of the resistors labeled $R1$, $R2$, and $R3$ and the supply voltage (Vcc) determine the upper and lower voltage limits. The formula for finding the lower voltage limit is

$$\text{Lower limit} = Vcc \frac{R3}{R1 + R2 + R3}$$

In a similar fashion, the value of the upper voltage limit is given by

$$\text{Upper limit} = Vcc \frac{R2 + R3}{R1 + R2 + R3}$$

The square wave output switches between a maximum value equal to the supply voltage and a minimum value of zero. The output frequency depends upon the values of the capacitor C (0.01 μF in Figure 4) and resistor $R4$ (33 K in the figure). The formula for finding the output frequency is

$$\text{Output frequency} = \frac{0.72}{CR4}$$

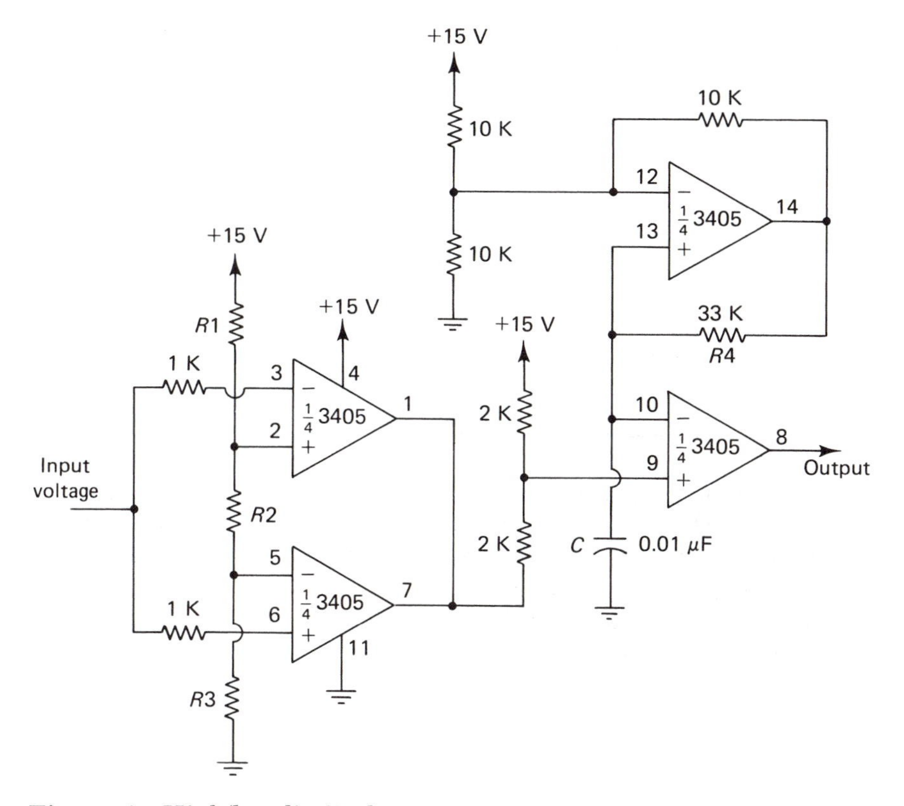

Figure 4 High/low limit alarm

In Figure 4, the values of C and $R4$ produce an output frequency of 2,200 Hz.

This circuit makes use of an interesting device from Motorola known as the MC3405. This chip contains two operational amplifiers along with two comparators. The op amps are similar in performance to those of the MC3403 family, while the comparators are much like the LM339 comparator ICs. The MC3405 can operate from a single DC supply voltage of 3 to 36 V or from dual voltages between 1.5 and 18 V. Thus, it is possible to vary the upper and lower voltage limits by changing the supply voltage while leaving the output frequency constant.

Alarm, overvoltage

A sudden surge in voltage from a power supply can damage many types of electronic devices and components. The usual approach to dealing with this problem is to incorporate a fuse in the line between the power supply and the circuit. But we may want an indication of when the voltage is starting to rise before it increases to a level which blows the fuse. The circuit in Figure 5 is designed for use

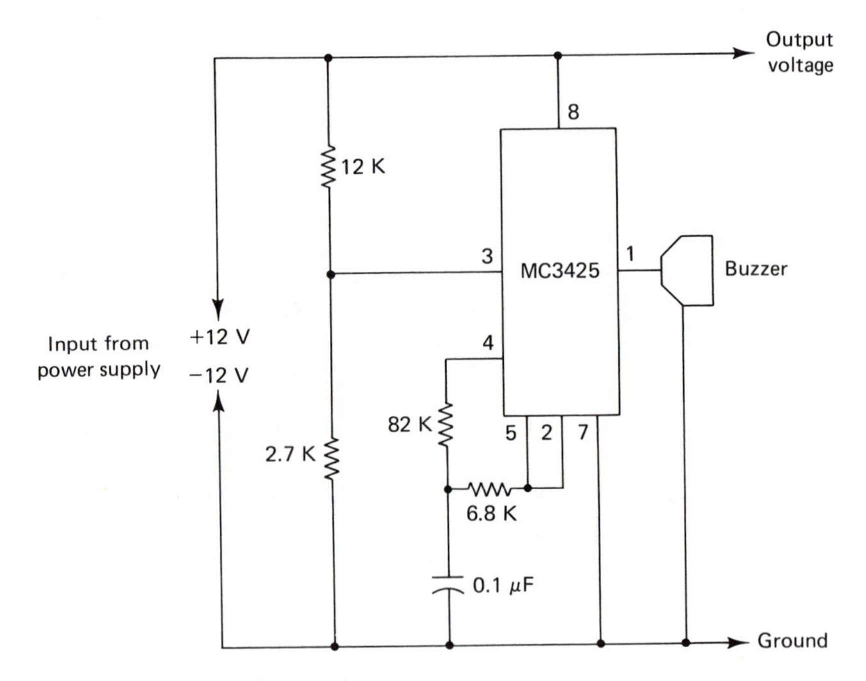

Figure 5 Overvoltage alarm

with power supplies rated at a nominal 12 V. If the output voltage rises to 13.6 V or more, the piezo buzzer will sound.

This circuit is built around an MC3425 power supply supervisory IC. Pin 3 of this device is an overvoltage sensing input connected to the output voltage, while pin 4 is an undervoltage sensing input connected to ground. The signals of the overvoltage and undervoltage inputs are applied to a comparator. The MC3425 also includes an internal 2.5 V voltage reference. The output of the voltage reference is applied as one input of a comparator, while the output of the overvoltage/undervoltage sensing comparator is applied to the other input. If this comparator is switched on, an output signal appears at pin 1 sufficient to drive most piezo buzzers of approximately 100 Ω. This device can be used with voltages of up to 40 V.

While this circuit can be very useful, it is wise not to rely totally upon it for protection of another circuit, since a sudden, catastrophic failure in the power supply could deliver a destructive voltage to a circuit before the power supply could be switched off. Thus, a fuse should still be used.

Amplifier, absolute value

An absolute value amplifier generates a positive output voltage for either polarity of an input signal, making it especially useful with input signals with a positive or negative peak greater than the other half of the signal cycle. For positive signals it acts as a noninverting amplifier, and for negative signals it functions as an inverting amplifier. Figure 6 shows such a circuit. The op amp used, an SE/NE5535 by Signetics, was selected because it has a high slew rate, which allows for more accurate detection of the maximum value. However, other op amps such as the 741 and 1458 may be substituted if lower accuracy can be tolerated. The diodes used may be general-purpose switching types such as the 1N914 or some equivalent.

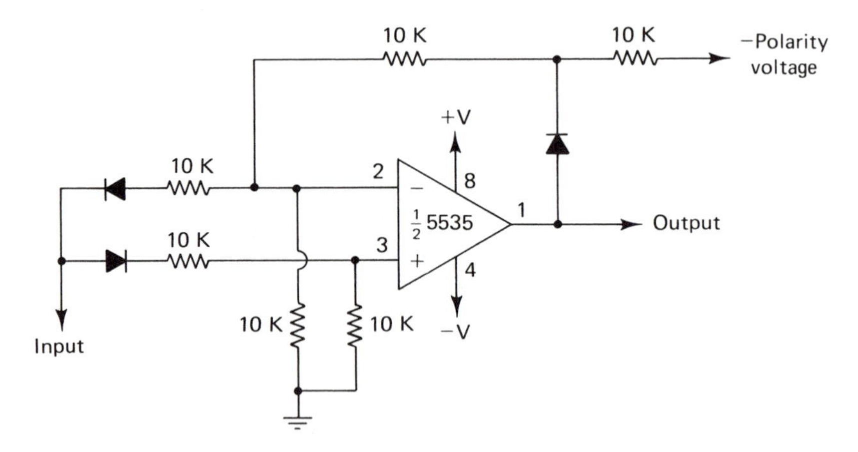

Figure 6 Absolute value amplifier

Amplifiers, audio

While general-purpose op amps may be used for audio amplification, several op amps optimized for that purpose have been developed and are generally preferable to "garden variety" op amps. Such devices have maximum gain and response at audio frequencies with low noise.

One of the first op amps designed for audio amplification was the 383, which was intended for automotive sound systems. This device can operate from a supply voltage in the range of +5 to +20 V, and it delivers 8 W at 3.5 A output current into a 4 Ω load. An 8 W audio amplifier using the 383 is shown in Figure 7.

An improved audio amplifier IC is the 386, which was designed for low-voltage consumer applications. This device can operate from 4 to 12 V, while consuming only 18 mW of current at 6 V in the

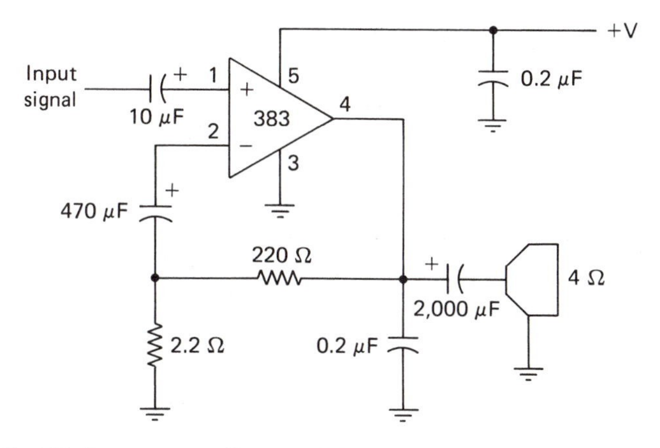

Figure 7 Eight-watt audio amplifier

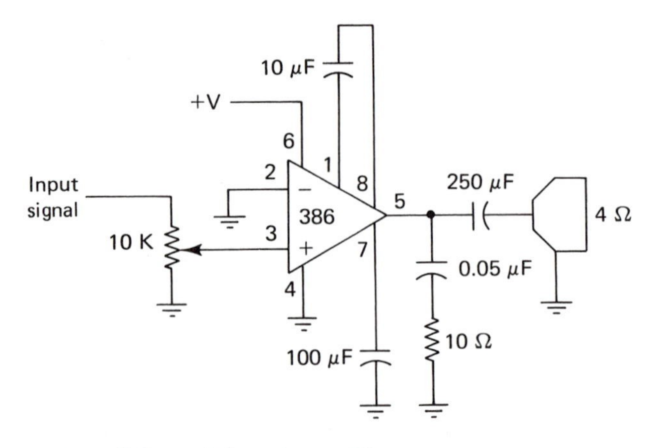

Figure 8 Amplifier with gain = 50

quiescent mode. The circuit in Figure 8 shows the 386 configured to deliver a gain of 50 into a 4 Ω load. The gain of the 386 is controlled by components connected between pins 1 and 8. If a 1.2 K resistor is added between pin 1 and the 10 μF capacitor, the gain is

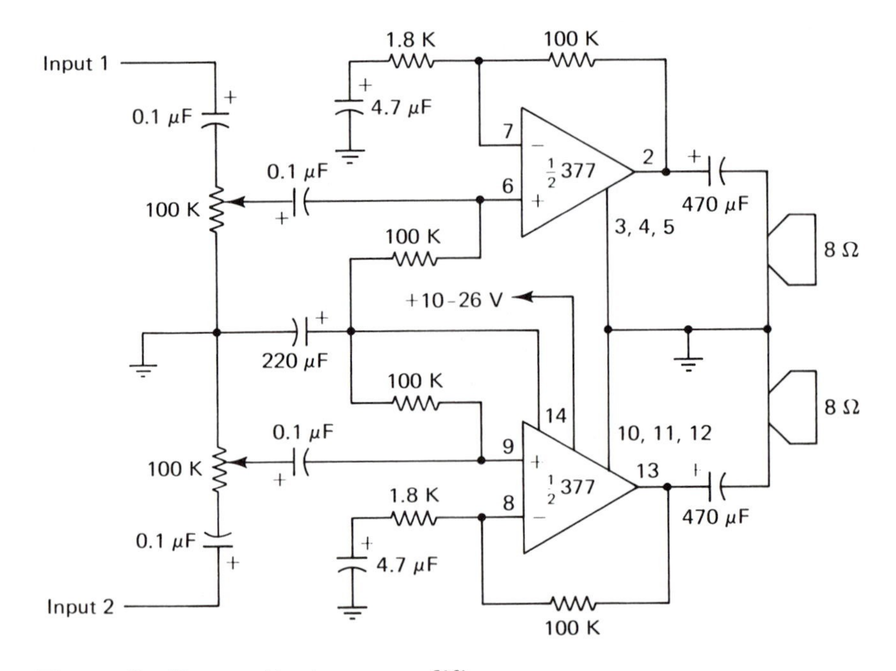

Figure 9 Two-watt stereo amplifier

increased to 200. If no components are connected between or to pins 1 and 8, the amplifier gain is 20.

Two-channel stereo amplifier systems are usually little more than two separate single-channel systems sharing a common power supply and ground reference point. Figure 9 is an example of such a system using both halves of a 377 dual op amp IC. Other dual op amps may be substituted, or two single op amps may be used. The output level of each channel is set by the two 100 K potentiometers. The maximum output of the circuit when the 377 is used is 2 W per channel.

Amplifier, audio distribution

An audio distribution amplifier is used to split an audio input signal into several separate but identical output signals. Some degree of amplification is necessary when this is done; otherwise, the signal loss can be serious enough to render the output signals unusable.

Figure 10 shows a three-output audio distribution amplifier

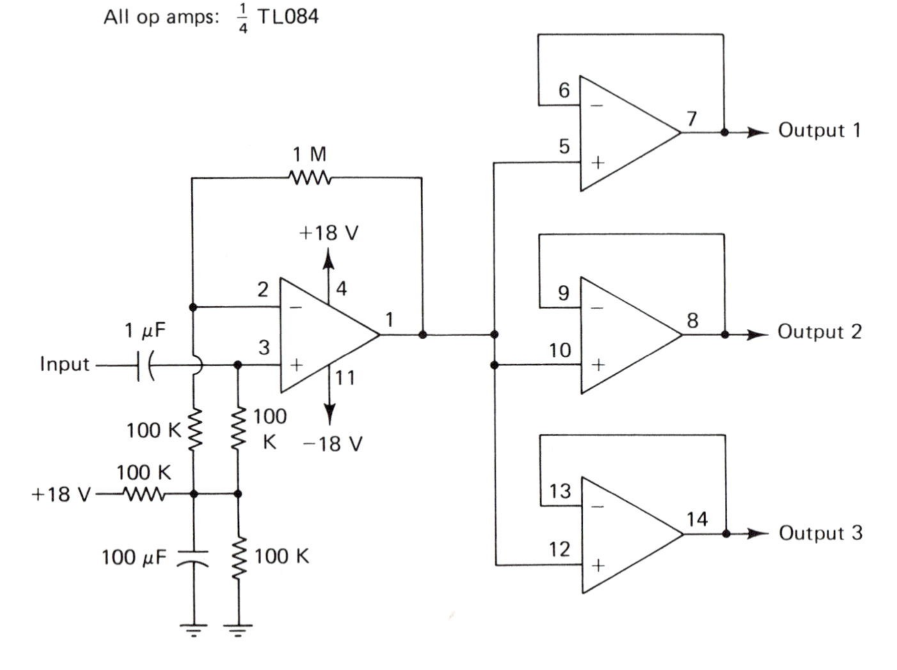

Figure 10 Audio distribution amplifier

which uses a TL084 quad JFET operational amplifier. This device has a high-input impedance, so it draws little current from the input signal source. The circuit basically offers unity gain and is designed for minimum distortion so that the output signal is a reproduction of the input signal.

Amplifier, audio with squelch

A *squelch* circuit is frequently used in radio communication receivers to turn off the audio output of the receiver until a signal equal to or greater than a certain level is received. This means that a receiver can be continuously monitored without having to listen to annoying background noise during periods when no signal is being received; also, signals too weak for successful communications are suppressed by the squelch.

Most squelch circuits are separate from the receiver audio amplifier stage. However, the circuit in Figure 11 combines both functions. The gain of the audio amplifier section is controlled by the

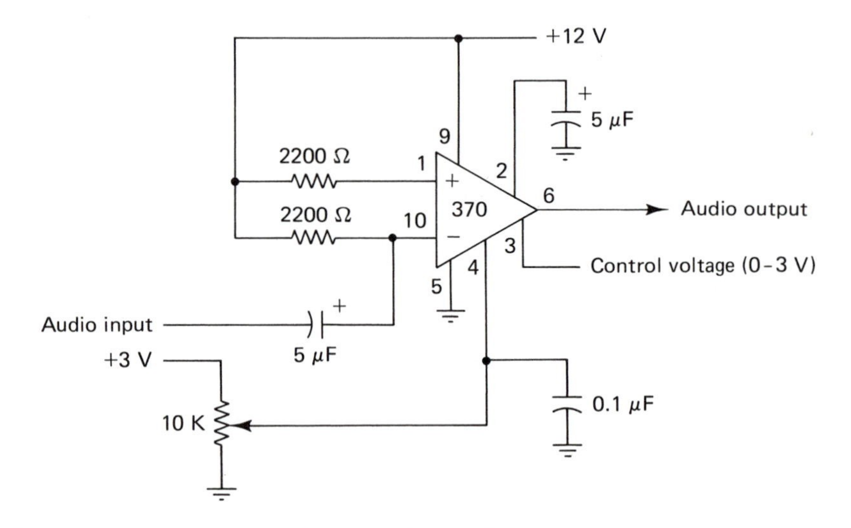

Figure 11 Audio amplifier with squelch

setting of the 10 K potentiometer, while the squelch function is set by a control voltage applied to pin 3 of the LM370 IC. The gain of this circuit as an audio amplifier is approximately 40 dB, as long as the control voltage applied to pin 3 is 0 V. If 3 V is applied to pin 3, the audio gain is attenuated by about 70 dB. By varying the control voltage between 0 and 3 V, a squelch cutoff point can be set so that an input signal must be above a certain level to produce an output from the stage. Note that the control voltage has no effect on the audio output level; this is still controlled by the setting of the 10 K potentiometer. The control voltage sets the input level required to produce a given audio output.

The utility of this circuit is not restricted to radio receivers. It can be used whenever it is desired to restrict the output of an audio amplifier to input signals of a given level or greater.

Amplifiers, buffer

Buffer amplifiers are also known as *unity gain* amplifiers. Both names are accurate, since the output of such a circuit is equal to the input. The purpose of a buffer is to isolate different sections or stages of a circuit from each other to prevent unwanted interaction between them.

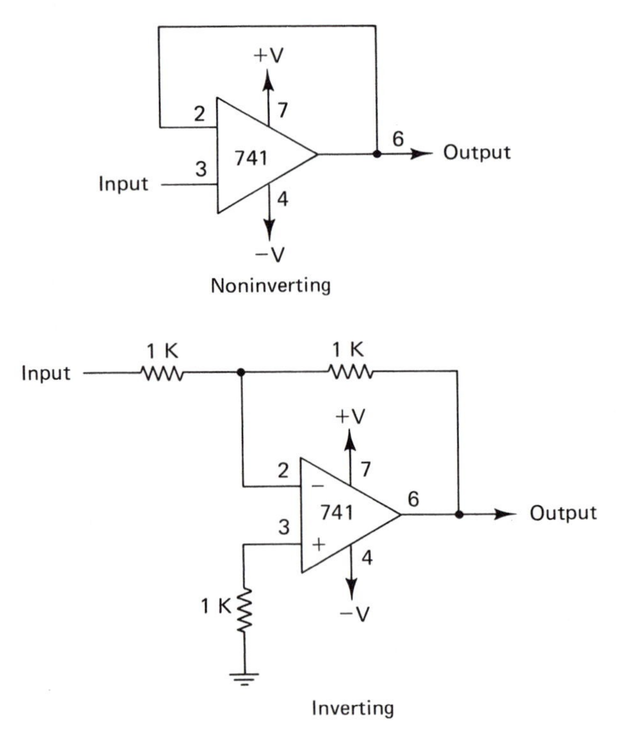

Figure 12 Buffer amplifier

Figure 12 shows noninverting and inverting versions of a buffer amplifier built around a 741. The noninverting buffer has an output signal with the same polarity as the input signal. In the inverting buffer, the output polarity is the opposite of the input. Other op amps can be used as long as the gain provided is equal to unity.

Amplifiers, DC servomotor

A servomotor is a motor whose rotation or speed is controlled by a corrective voltage fed into the motor circuit. Servomotors have been used for years in industrial applications and are now becoming more frequent in robotics. Substantial output current is necessary in servomotor control applications, however, but it can be provided by a power operational amplifier. Figure 13 shows one design which uses the 759 power op amp. This particular device provides up to 325 mA of output current depending upon the control voltage. Similar power op amps may be used depending upon the current demands of the servomotor being controlled. This circuit can also be used for other applications where a great deal of output current is needed.

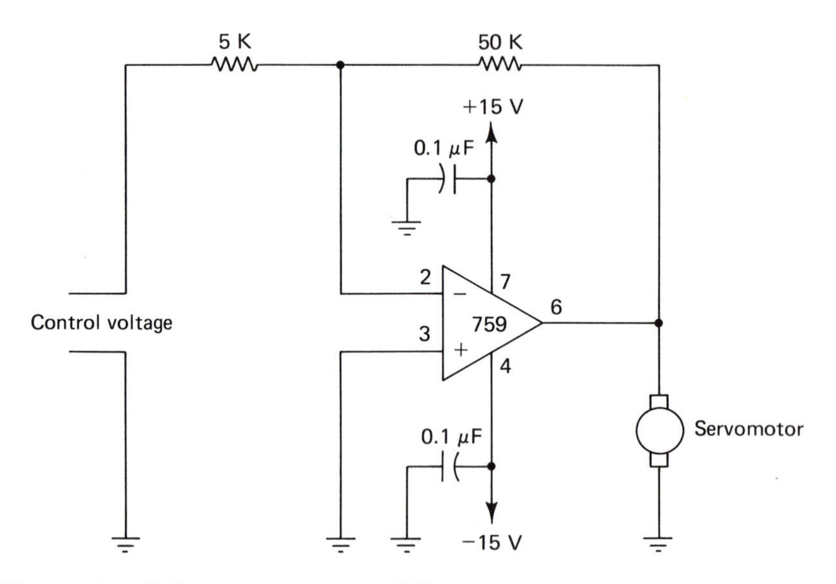

Figure 13 DC servomotor amplifier

Amplifiers, difference

The operational amplifier circuit was originally developed for use in analog computing systems. In World War II, several analog computers, using operational amplifiers constructed with vacuum tubes, were built and placed into service for aircraft and guided missile design. Today's op amp ICs can be used for the same purposes the original op amp circuits were designed for.

The output voltage of the circuit in Figure 14 is given by

Output voltage = input voltage 2 − input voltage 1

The output polarity of the amplifier in the figure is normally equal to that of the input voltages. However, there are some cases when an inverted output is desired. Figure 15 shows a circuit that provides this. The output voltage is computed in the same way as the circuit in Figure 14, but the result is negated.

In both circuits, the input voltages must not exceed the supply voltages; the output voltage cannot exceed the supply voltage. The

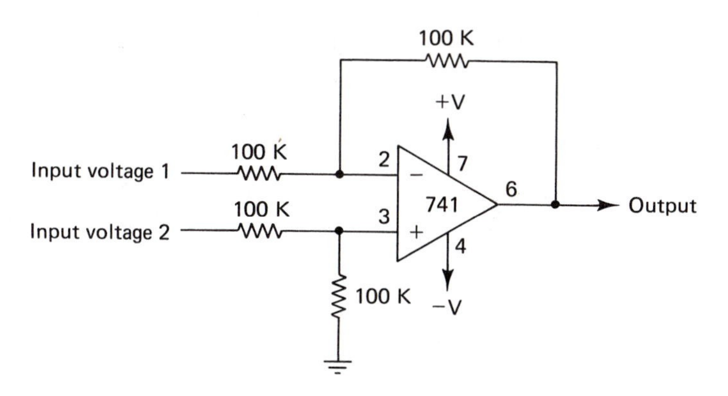

Figure 14 Difference amplifier

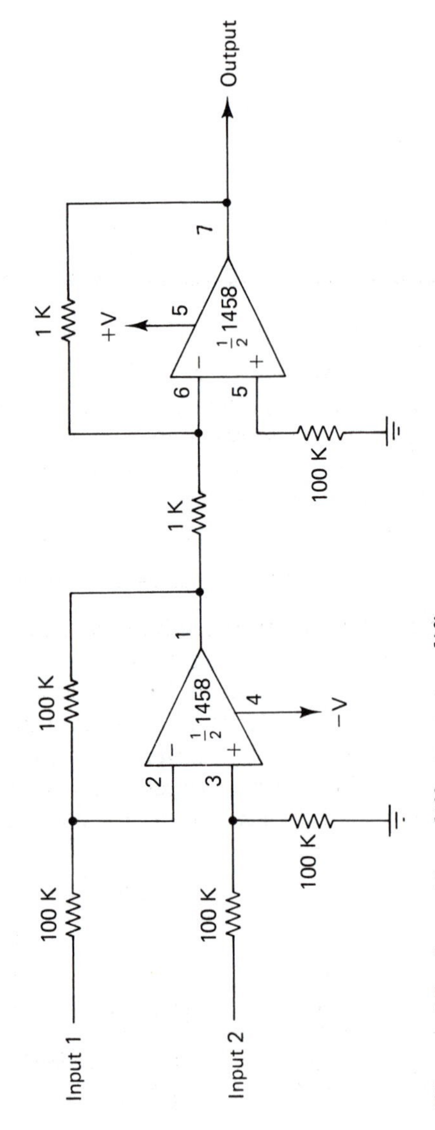

Figure 15 Inverting difference amplifier

accuracy of the output depends upon the tolerance of the resistors used; 1 percent resistors will give a reasonably accurate result, while 20 percent resistors will be much less accurate.

In actual analog computers, difference amplifiers are used for both subtraction and division. Division is performed by repeated subtraction.

Amplifier, instrumentation

The requirements for amplifying the output of a measuring instrument are more critical than those of many other applications. High gain is normally necessary, often on the order of 1 million or greater, due to the low output of many sensors and other measuring devices. The circuit should have a high input impedance in order to avoid producing inaccurate readings due to drawing too much current from the measuring device, but it also must be highly sensitive to faint input signals. Also, it must have low drift and adequate output current. Finally, it should be well isolated from the input circuit. These requirements cannot be satisfied by most common op amp ICs, so several precision op amps designed specifically for use in instrumentation amplifiers have been developed. Figure 16 shows one such device, the LT1024 dual-precision op amp, used for that purpose.

The supply voltage requirements for the circuit are ±20 V. With the resistor values shown, the circuit has a gain of 100.

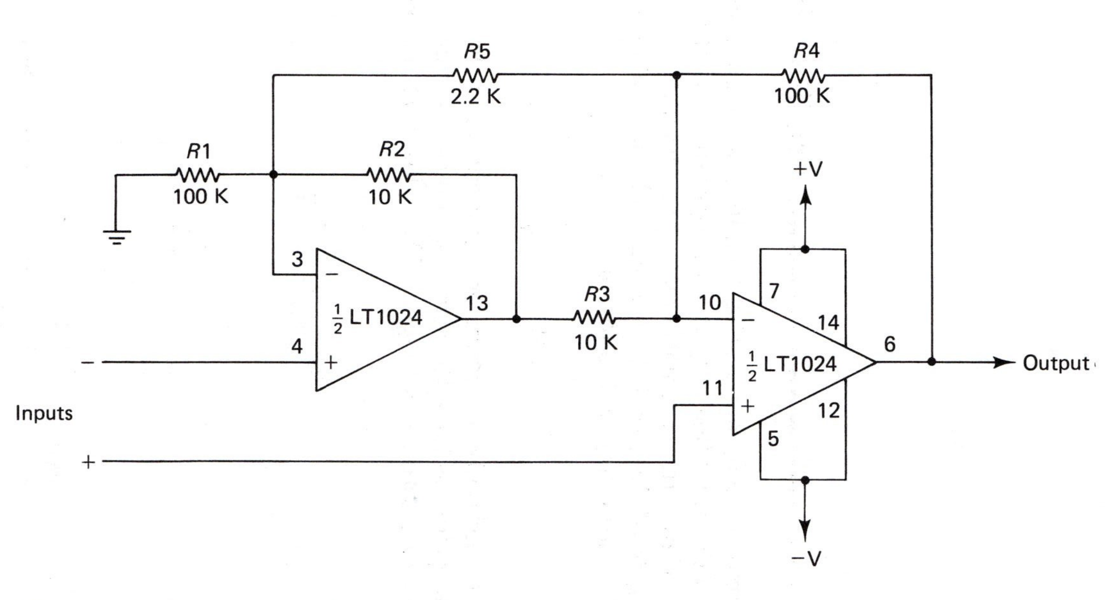

Figure 16 Instrumentation amplifier

Amplifier, inverting

This common op amp application amplifies an input signal, but the polarity of the amplified output is the opposite of that of the input. Figure 17 shows the schematic for such a circuit.

The gain of the circuit is determined by the values of resistors $R1$ and $R2$, according to the formula

$$\text{gain} = -\frac{R2}{R1}$$

Resistor $R3$ grounds the noninverting input and helps ensure that both op amp inputs have the same resistance and that a zero input signal causes a zero output signal. The value of $R3$ is found by the formula

$$R3 = \frac{R1R2}{R1 + R2}$$

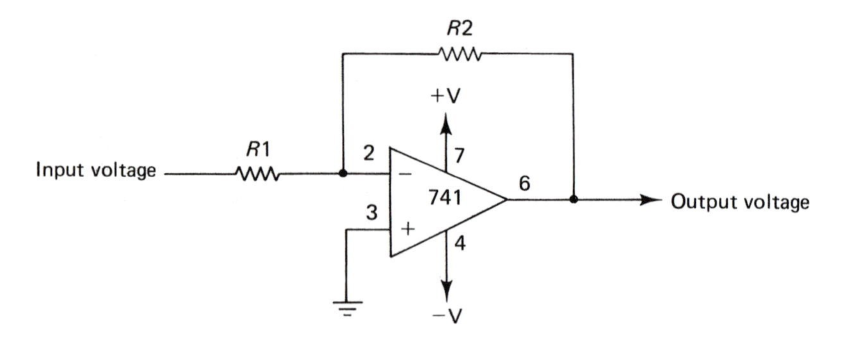

Figure 17 Inverting amplifier

*R*3 may not be necessary in all cases. If so, the noninverting input can be connected directly to ground. If the circuit fails to function properly without it, however, try adding the correct-valued *R*3.

Amplifier, noninverting

The noninverting amplifier is similar to the inverting amplifier. The only real difference is that the output and input have the same polarity. Figure 18 shows the schematic for this type of amplifier.

The gain of this circuit is found in a similar fashion to that of the inverting amplifier:

$$\text{gain} = 1 + \frac{R2}{R1}$$

The circuit is best suited for use with DC signals. If AC signals are used as the input, it is best to set the gain to unity for a DC input. This is done by placing a capacitor between $R1$ and ground. The exact value is large, usually 4.7 μF or greater, and is determined experimentally by substituting values until the circuit performs with best results.

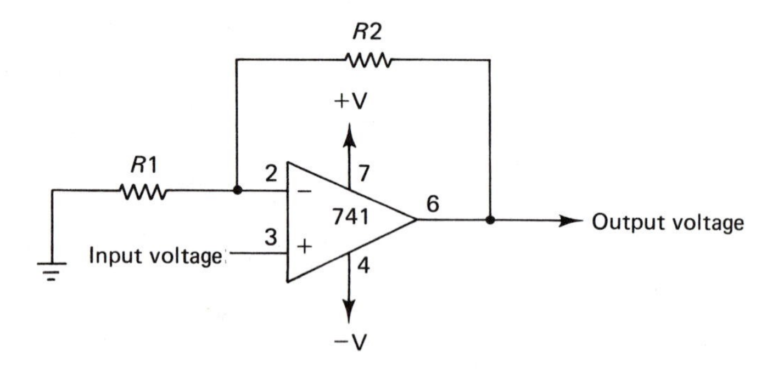

Figure 18 Noninverting amplifier

Amplifier, programmable gain

Since the gain of both the inverting and the noninverting amplifier depends upon the values of two resistances, it is controllable by switching in a different value of resistance. This can be done by a mechanical switch or relay, but a better approach for many applications is to use an analog switch IC instead. Figure 19 shows such a circuit based upon the Intersil IH5009 and IH5010 analog switch devices. The ICL8007 operational amplifier is a JFET input device, but most other op amps will work as well as long as the different input impedances are kept in mind.

The IH5009 and IH5010 are identical in function and other characteristics, except for the input signal levels applied to them. The IH5009's inputs will be activated when a signal in excess of +15 V is applied to one of them, with 15 V being recommended as an input signal voltage. The IH5010 is designed to accept TTL (5 V) input levels. Only one input signal can be applied at a time to either device, and neither requires an external supply voltage.

Operation of this circuit is easily understood from the schematic. If a voltage is present at pin 1 (labeled "10 K"), then the 10 K resistor connected to pin 2 is switched into the circuit. The same procedure is followed with the other three resistors. If no control voltage is present at any of the input pins, then none of the resistors is switched into the amplifier circuit.

This approach to selecting the gain of an amplifier is much more efficient in terms of current consumption than relays, and is also much faster. The configuration can be adapted to other circuits where it is desired that different component values be switched in using control voltages. One change you might want to make is to include a mechanical switch to remove the IH5009/5010 and its re-

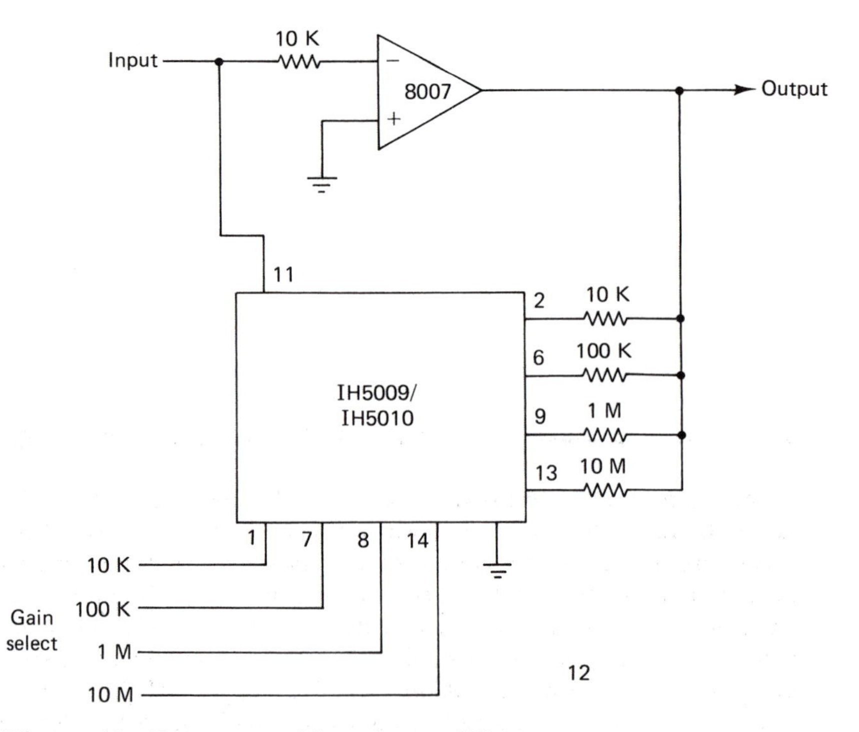

Figure 19 Programmable gain amplifier

sistors from the amplifier circuit and replace it with a single fixed resistor when no control voltages are to be present at the IH5009/IH5010 inputs.

Amplifier, RIAA standard

The Recording Industry Association of America (RIAA) developed a standard several years ago for phonograph and record player playback to prevent large undulations from breaking through phonograph record groove walls. This standard calls for low frequencies to be attenuated and high frequencies to be boosted in the manufacture of phonograph records. This means that on playback an amplifier must have an *inverse* frequency response for normal sound. Figure 20 shows an amplifier which meets RIAA standards built around a 380 audio amplifier IC. The input of the amplifier is designed to be from a crystal cartridge or similar high-impedance source. The 2.7 Ω resistor and 0.1 μF capacitor between the 380's output and ground help stabilize the amplifier under high current loads; they may be omitted if such loads will not be demanded. Output power is approximately 2 W.

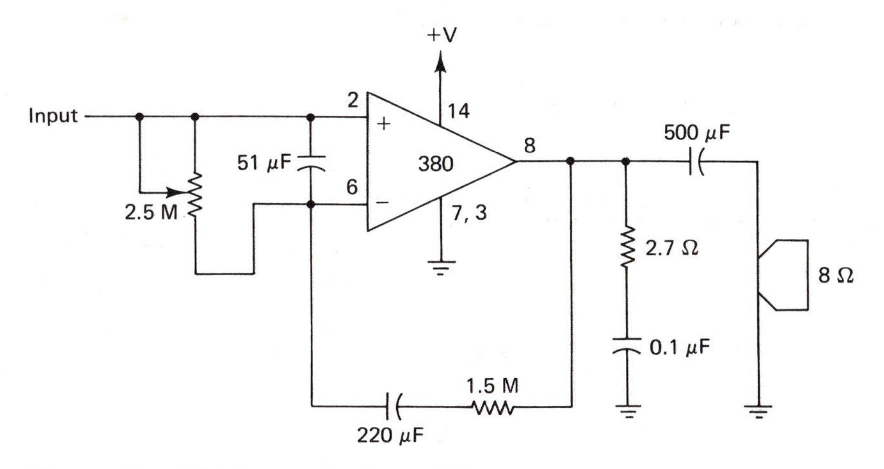

Figure 20 RIAA standard amplifier

Amplifiers, summing

As mentioned earlier, operational amplifiers were originally designed for analog computers. The circuit in Figure 21 adds two input voltages together to produce a single voltage equal to their sum.

The most important consideration to remember with this circuit is that its output cannot exceed the supply voltage. The output voltage is given by the formula

$$\text{Output voltage} = -(\text{input voltage 1} + \text{input voltage 2})$$

The circuit is an inverting summing amplifier, since the output is the reverse polarity of the sum of the input voltages.

The circuit in Figure 22 shows a noninverting summing amplifier built using both halves of a dual op amp IC. The first op amp section is the actual summing amplifier, while the second op amp is a unity-gain inverting amplifier.

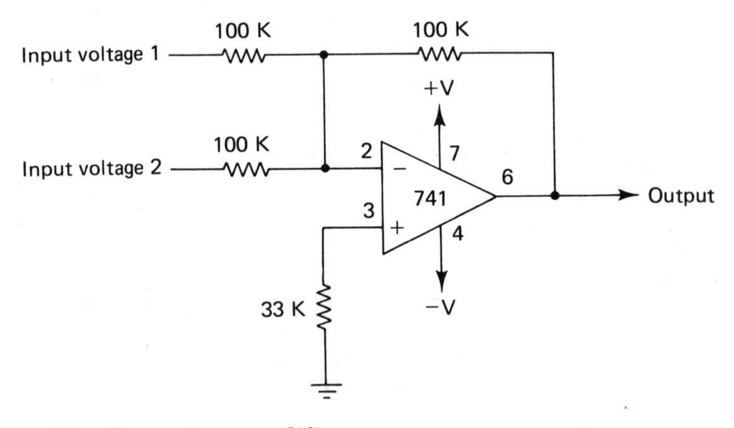

Figure 21 Summing amplifier

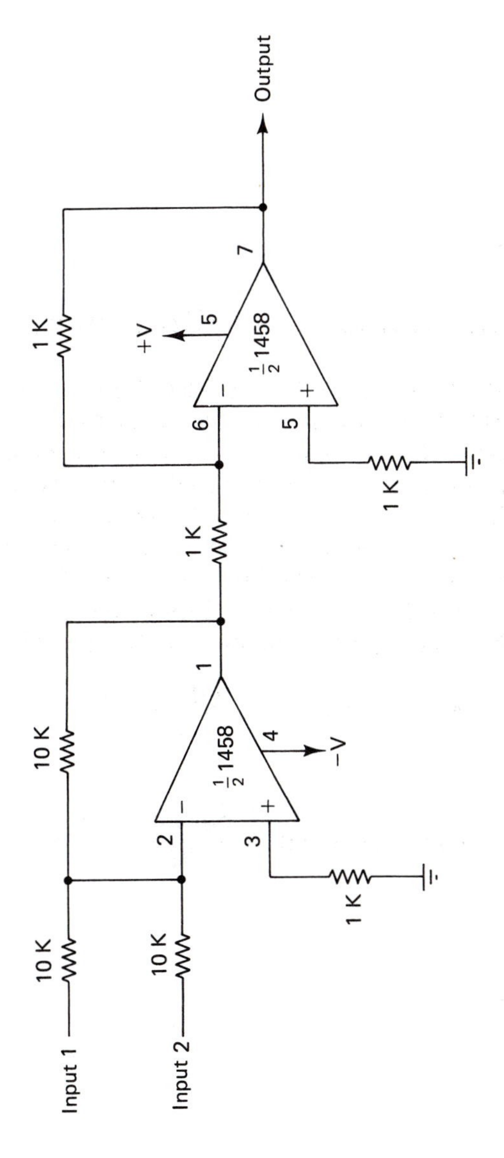

Figure 22 Noninverting summing amplifier

Amplifier, transconductance

The transconductance amplifier is also known as a *voltage-to-current converter.* This circuit produces a current across a load resistance, labeled "*RL*" in Figure 23, which is proportional to the input voltage. The load resistance is seldom an actual resistor; usually it is another device, such as a meter or measuring device. The output current is "sensed" by resistor *RS*. The resulting voltage is then fed back in series with the input voltage. The output current is determined either by dividing the input voltage by *RS* or by dividing the output voltage by the sum of *RL* and *RS.* That is,

$$\text{Output current} = \frac{\text{input voltage}}{RS}$$

or

$$\text{Output current} = \frac{\text{output voltage}}{RL + RS}$$

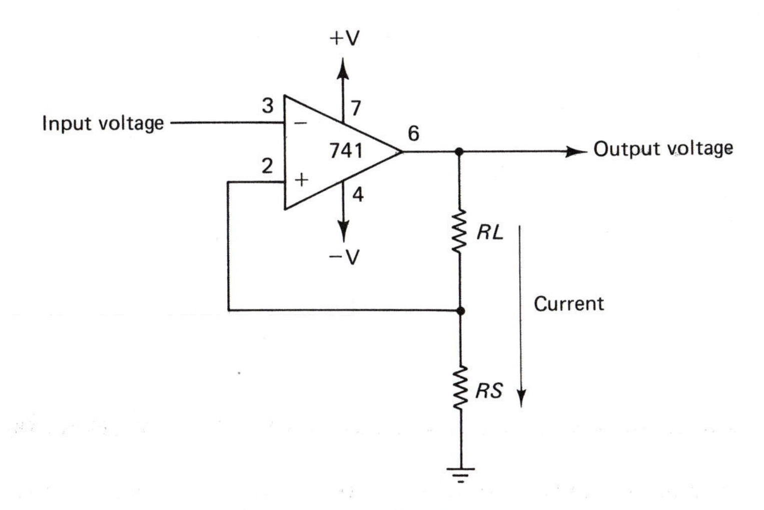

Figure 23 Transconductance amplifier

Amplifier, transresistance

A transresistance amplifier is also known as a *current-to-voltage converter.* The circuit shown in Figure 24 produces an output voltage that is proportional to an input current. The output voltage is determined by the formula

$$\text{Output voltage} = -\,(\text{input current} \times R)$$

The gain of this circuit is the output voltage divided by the input current; it can also be found simply by negating the value of R. (A value of 1 K for R translates to a gain of $-1{,}000$.)

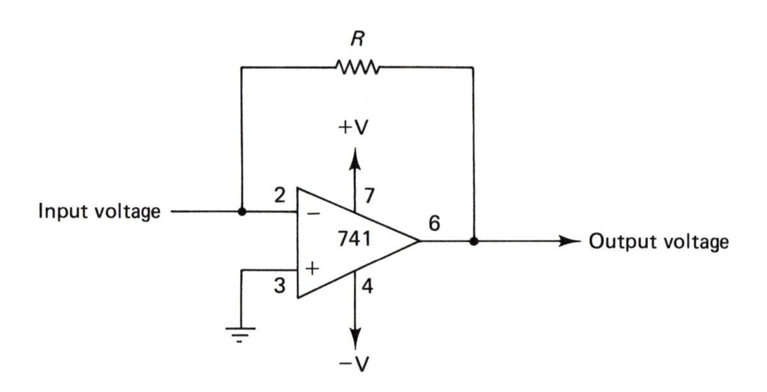

Figure 24 Transresistance amplifier

Amplifier, video

Video amplifiers require a wide bandwidth gain (due to the nature of video signals) at frequencies well into the VHF range. Figure 25 shows a circuit which offers a gain of approximately 35 dB at 100 MHz with a bandwidth adequate for video signals. The circuit uses an MC1590G wideband RF/audio amplifier IC; the supply voltage

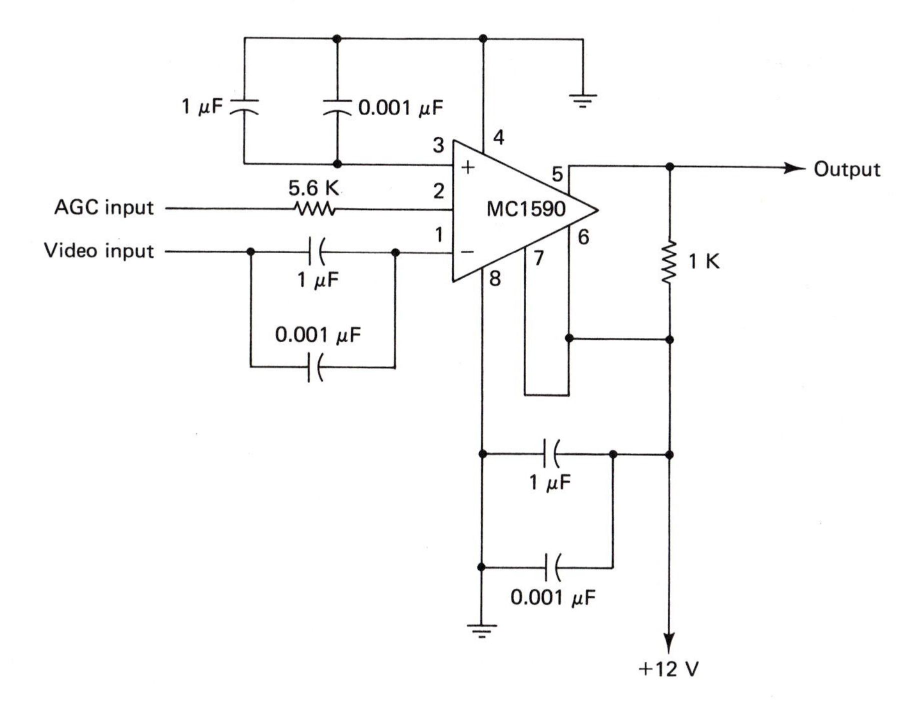

Figure 25 Video amplifier

can range from +6 to +15 V. The gain of the amplifier can be controlled by the AGC voltage, which must be equal to or less than the supply voltage. Additional gain can be obtained by cascading several of these circuits. As with all circuits designed to operate at VHF, component placement and lead length are crucial.

Amplifier, 30 MHz

The MC1590 can be used as a broadband amplifier at 30 MHz for applications such as an IF or RF amplifier, as shown in Figure 26. In this configuration, the gain is over 50 dB at 30 MHz with a bandwidth of over 1 MHz. The gain can be controlled through a voltage applied to pin 2 through a 5.6 K resistor; the greater the voltage applied to pin 2 (up to the supply voltage), the lower the gain of the circuit. Inductor *L*1 is 12 turns of #22 wire wound on a T37-6 toroid core or some equivalent. Transformer *T*1 has a primary of 17 turns of #20 wire on a T44-6 toroid core with a secondary of two turns of #20 wire. This circuit must not be allowed to radiate into an antenna without an appropriate FCC license.

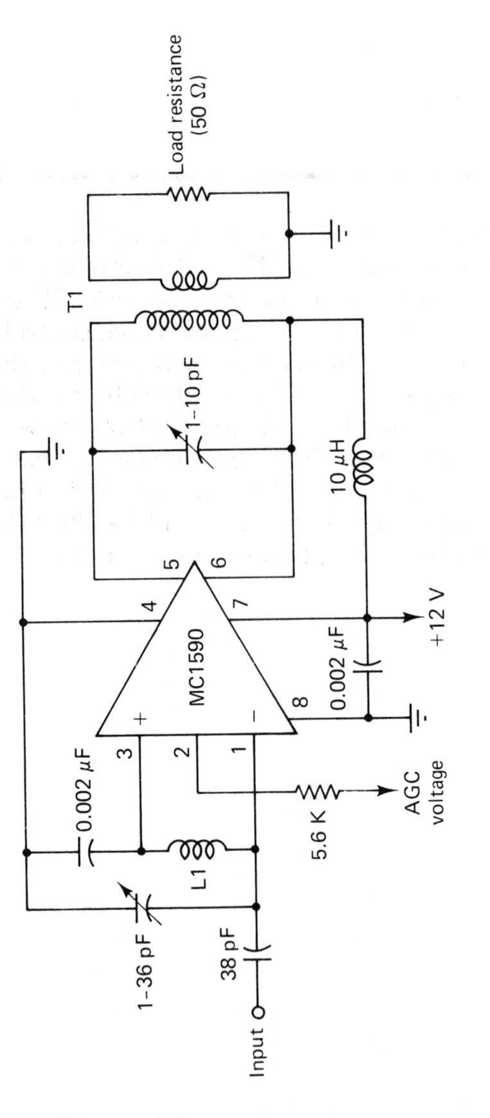

Figure 26 30 MHz amplifier

Amplifier, 30 to 900 MHz

The circuit shown in Figure 27 is a broadband linear RF amplifier capable of covering 30 to 900 MHz and delivering approximately 10 dB of gain throughout this range. This circuit was originally designed for use in cable TV amplifier systems and has a very low

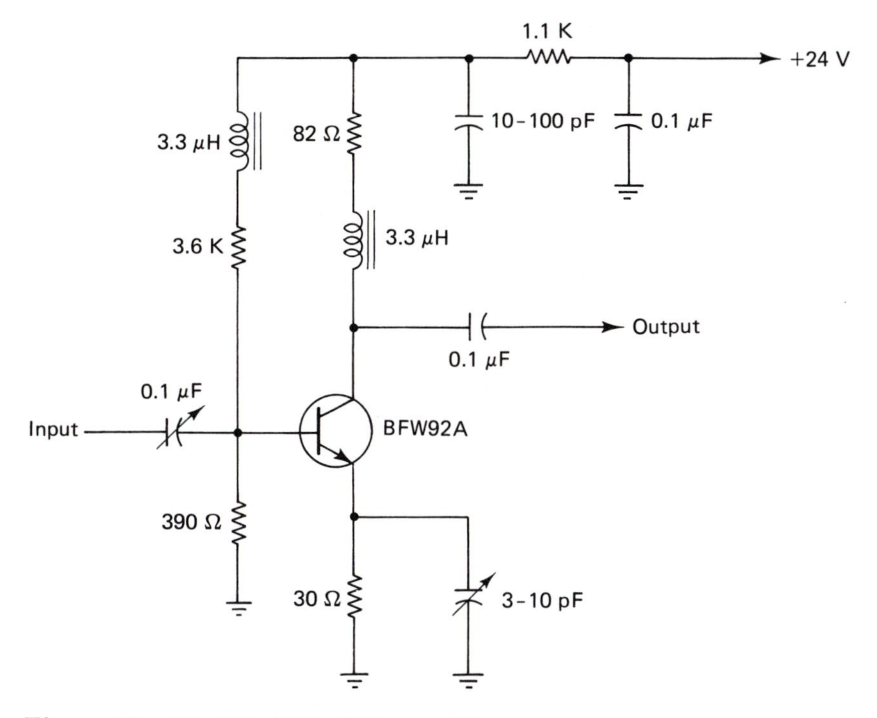

Figure 27 30–900 MHz RF amplifier

noise figure (typically 2.7 dB at 500 MHz). The two variable capacitors (10–100 pF and 3–10 pF) both adjust for maximum output for a given input frequency. Input and output impedances are approximately 75 Ω. Without an appropriate FCC license (e.g., a Technician class amateur radio license), the circuit must not be allowed to radiate into an antenna.

As with all high-frequency RF circuits, construction of this circuit requires care. All fixed-value capacitors are "chip" types, while the two 3.3 μH inductors are of the molded variety. All part leads should be as short as possible, and a good RF ground is a necessity. At the upper frequency limits of this amplifier's operation, unwanted inductances can occur and produce improper operation. Check for proper operation by measuring the collector current and collector-emitter voltage (Vce). The nominal values should be 10 mA and 10 V.

Amplifier, 400 MHz

This circuit, shown in Figure 28, uses a 2N3866 transistor in a class C configuration to produce up to 1 W of output at 400 MHz. Inductor $L1$ consists of two turns of #18 wire wound $\frac{1}{4}$ inch in diameter and $\frac{1}{8}$ inch in length. $L2$ is a one-turn ferrite RF choke with an impedance of approximately 450 Ω. $L3$ is $2\frac{3}{4}$ turns of #18 wire wound $\frac{1}{4}$ inch in diameter and to a length of $\frac{3}{16}$ inch. The exact values of $C1$ and $C2$ will lie between 8 and 60 pF and should be determined experimentally during circuit testing for best performance. Thirty milliwatts of input power will be sufficient to drive this circuit to 1 W output at 400 MHz. As with all circuits operating at these frequencies, component placement and lead lengths are critical and a good RF ground is essential for proper operation. This is another circuit that should not be allowed to radiate into an antenna without an appropriate FCC license.

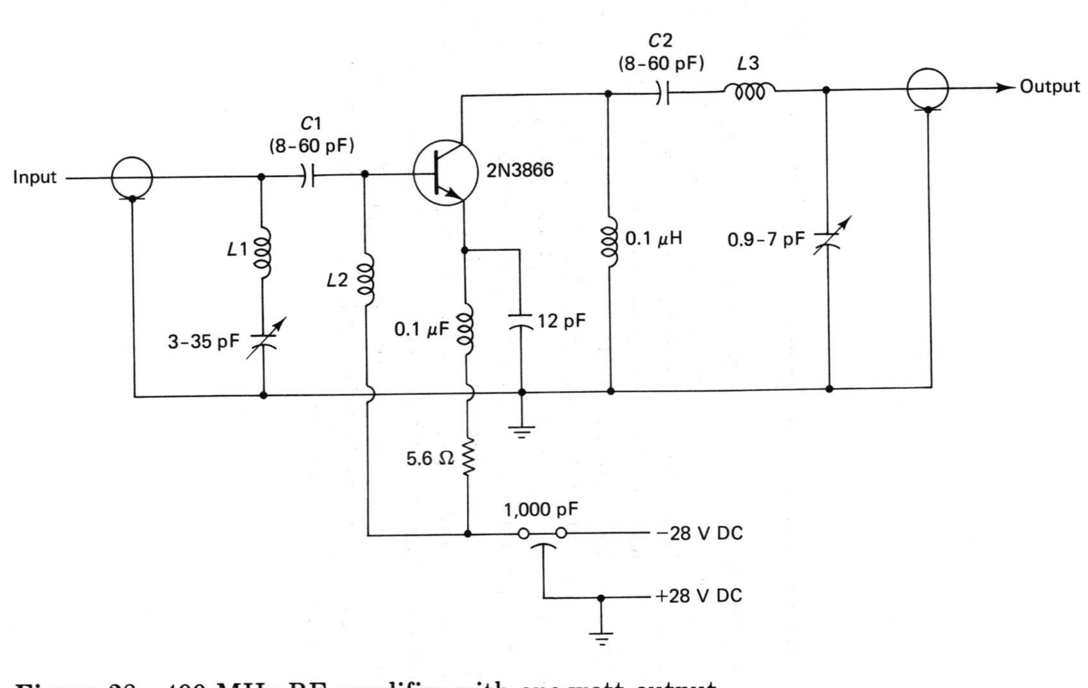

Figure 28 400 MHz RF amplifier with one-watt output

Amplifier, 500 MHz

An output of up to 400 mW at 500 MHz in class C operation is provided by the amplifier in Figure 29. Inductor *L*1 is a copper strip 1 inch long, $\frac{1}{4}$ inch wide, and 0.005 inch thick. *L*2 is a $\frac{3}{4}$ inch long length of #18 wire. *L*3 is formed from two turns of #18 wire wound $\frac{1}{4}$ inch in diameter. The RF bead is an Indiana General 56-590-65 or an equivalent. All variable capacitors are adjusted for maximum output on the desired frequency. As with all circuits operating at higher frequencies, component placement and lead lengths are crucial and a good RF ground is essential. The output of this circuit must not be radiated through an antenna without the appropriate FCC license.

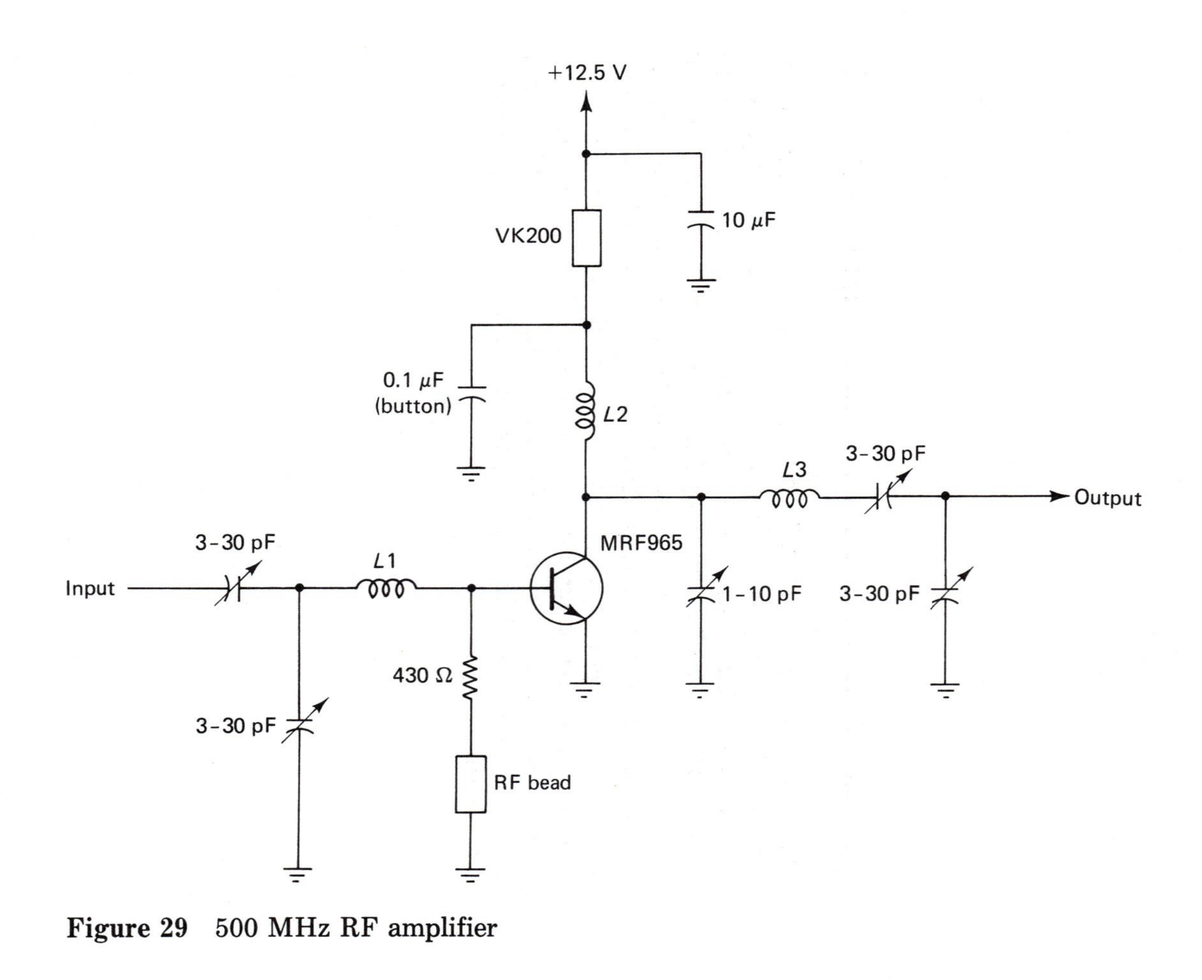

Figure 29 500 MHz RF amplifier

Attenuator, electronic

Most attenuator circuits in electronics are strictly resistive, passive devices which reduce the strength of signals applied to them by dissipating some of the input as heat. There are many situations in which this is not the best solution, however, and it is better to electronically attenuate a signal. Motorola has developed a device, known as the MC3340, specifically as an electronic attenuator. It can be thought of as something of an "inverse op amp" in that it

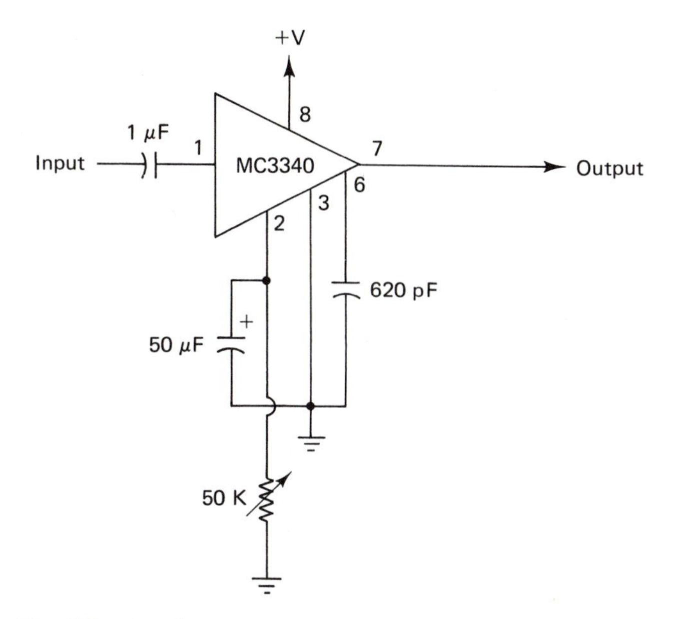

Figure 30 Electronic attenuator

can attenuate (as opposed to amplify) a signal in a linear manner. The degree of attenuation is determined by a control voltage applied to pin 2 of the MC3340. The supply voltage can range from 8 to 18 V, and the control voltage can range from 0 to slightly less than half the supply voltage (i.e., for a 12 V DC supply, the control voltage can be up to 5.5 V). The maximum current that can be sunk at pin 2 is 2 mA. Attenuation increases as the control voltage increases, up to a maximum of 90 dB.

The control voltage can be supplied through a potentiometer, as shown in Figure 30, which may be located some distance from the circuit. The frequency response of this circuit is flat up to about 1 MHz, while total harmonic distortion at maximum attenuation is 3 percent or less and is significantly lower at reduced attenuation.

Audio power meter

The LM3915 IC is capable of sensing analog voltage levels and driving 10 LEDs connected to its outputs. The exact number of LEDs lit up depends upon the level of the analog inputs. The current supplied at the LM3915's outputs is regulated, eliminating the need to use resistors with LEDs. Figure 31 shows the LM3915 configured to give a visual indication of the input signal power level by the number of LEDs lit up. An 8 Ω speaker is provided to allow monitoring of the audio input.

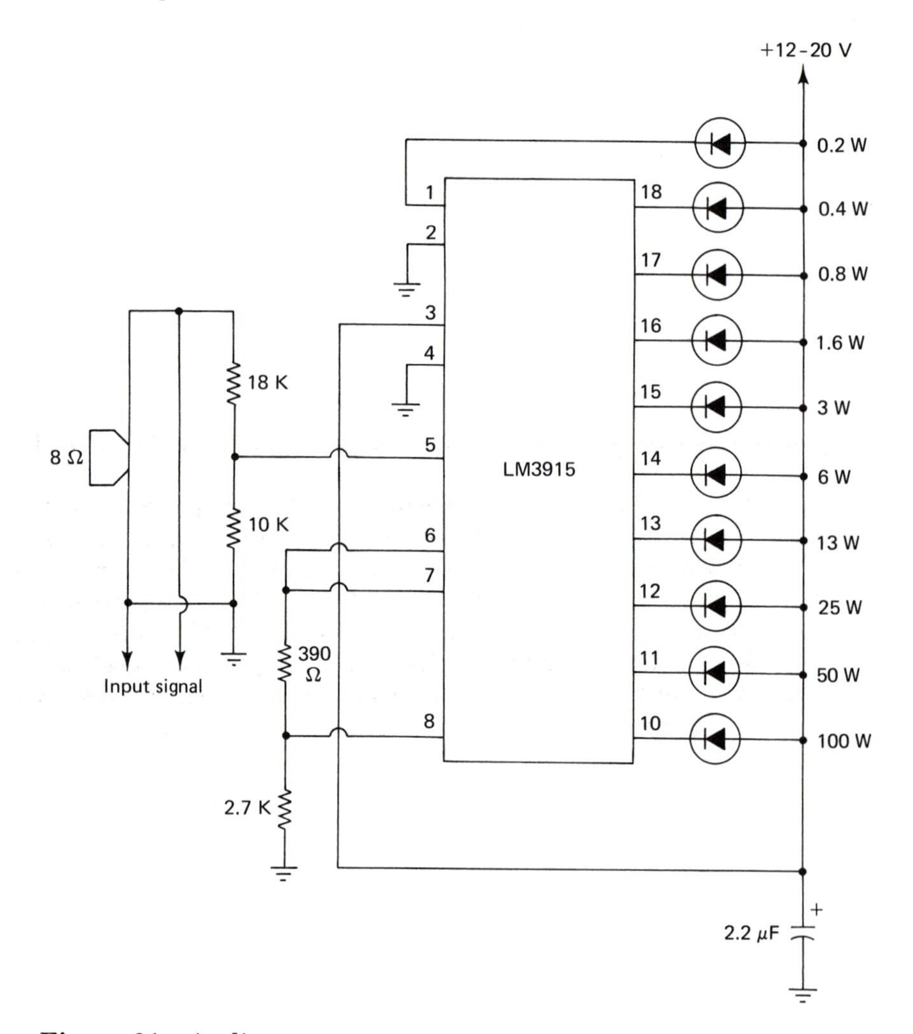

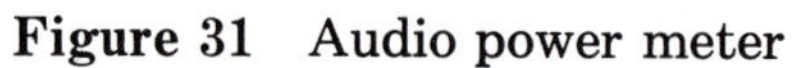
Figure 31 Audio power meter

b

Bass booster

Audio circuits (amplifiers and the like) are frequently more responsive to certain audio ranges, particularly high frequencies, than they are to others. This can result in audio signals with excessive treble and inadequate bass, producing a "tinny" sound. The circuit in Figure 32, a so-called bass booster which emphasizes the bass frequencies of an audio signal, is actually an audio amplifier which responds better to lower frequencies than it does to higher ones. The amount of bass emphasis is controlled by the 10 K potentiometer. The output power of the circuit is usually less than 0.5 W (depending upon the supply voltage used), and the output can drive an 8 Ω speaker or another amplifier stage that has a similar input impedance.

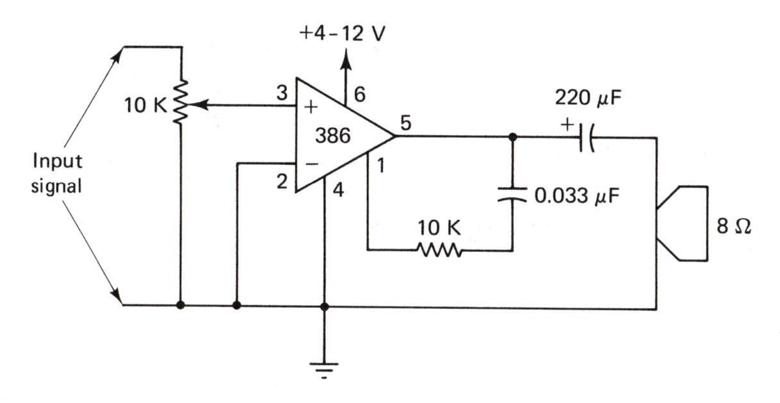

Figure 32 Bass booster

Battery charger

This circuit, shown in Figure 33, is designed to recharge 12 V batteries from an input voltage of 13.8 V or greater up to 35 V. The output voltage is held to a stable 12 V at output currents up to 1.5 A by the 317T voltage regulator IC. Capacitors are not needed since the 317T's ripple rejection is typically 80 dB. The circuit should be used only with those batteries, such as storage cells or ni-cads, which are intended to be recharged.

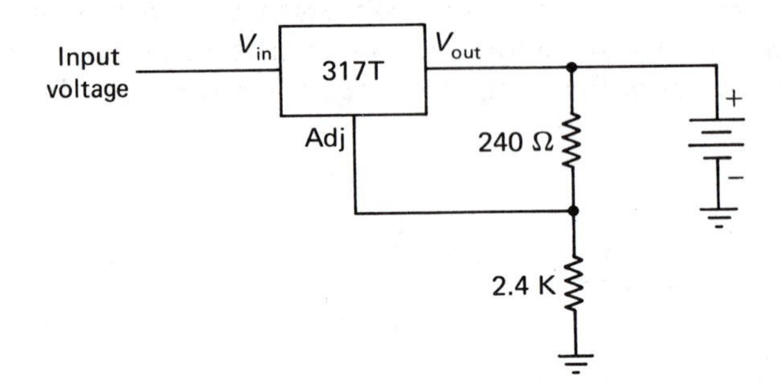

Figure 33 12 V battery charger

Bus buffer

This circuit serves as a "drawbridge" between two eight-bit data buses labeled Bus A and Bus B in Figure 34. The 74244 is a TTL noninverting octal buffer device whose operation is controlled by the enable signal applied to pins 1 and 19. When the enable signal is high, Buses A and B are isolated from each other and no data transfer can take place between them. But when the enable signal is low, data on Bus B can move to Bus A. However, the data on Bus A can never move to Bus B using this circuit. Since the 74244 is a noninverting device, the data moved from Bus B to Bus A does not change. If inversion is necessary, this configuration can be used with the 74240 inverting octal buffer. The pin numbers, connections, and operation remain the same.

Figure 35 shows a circuit with a similar function to that of Figure 34 designed for four-bit buses. The 4066 is a CMOS quad bilateral switch capable of handling analog and digital signals. The data on Bus B is isolated from Bus A as long as the enable signal is low (i.e., off). When the enable signal goes high, data can move from Bus B to Bus A.

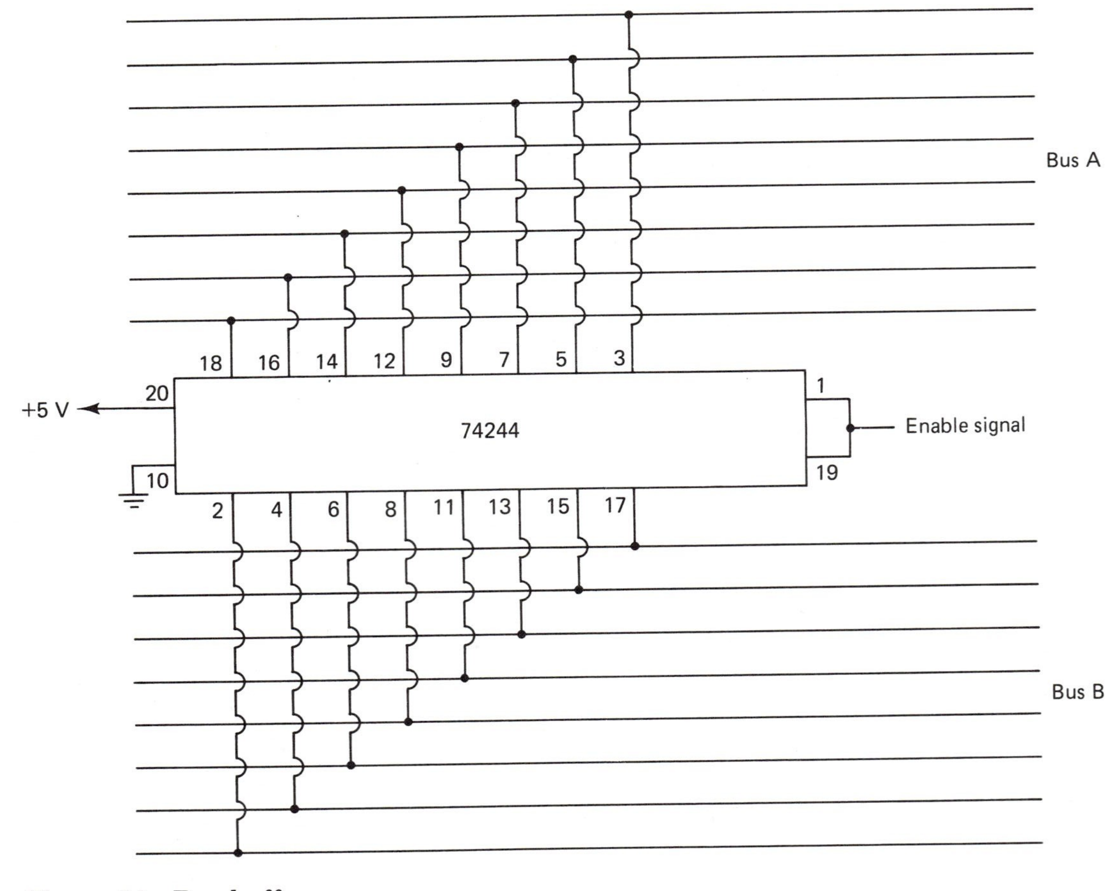

Figure 34 Bus buffer

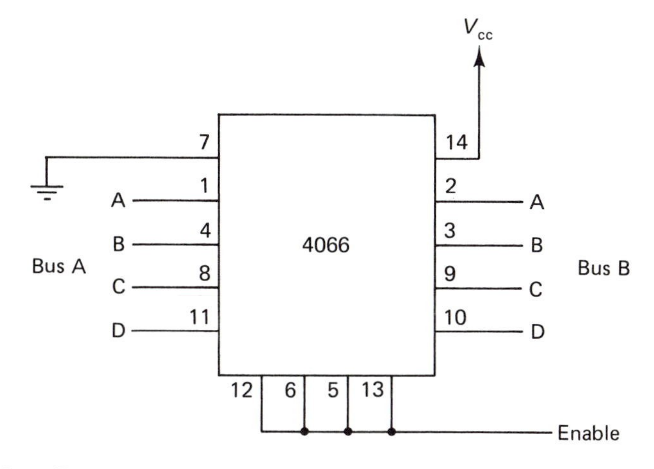

Figure 35 Bus control

Bus register

A bus register is a circuit which serves as a temporary storage area for data before or after transfer on the bus. Such circuits normally allow control over when data gets loaded into the register as well as when it is read out.

The circuit in Figure 36 uses a 74373 TTL octal D-type latch device to form a general-purpose eight-bit storage register. Two signals, an output control and an enable signal, control the operation of the circuit. If the output control signal is high, the enable signal and data inputs are overridden and the circuit is effectively disconnected from the bus. Thus, the output control signal must be low to load and transfer data to the bus. When it is, and when the enable signal is high, the signals appearing on the bus (the 74373's output) are the same as those at the device's inputs. When both output control and enable are low, no further input is loaded and the last set of output signals are "locked" as the output signals shown to the bus.

Another bus register configuration is shown in Figure 37. This configuration, called a common input/output register, is capable of reading and writing data to an eight-bit bus. The operation of the circuit is controlled by both the clock and output control signals. When the output control is high, the circuit will read data on the bus and load it into the 74374 on the rising edge of a clock pulse. When output control is low, the device reads data from the circuit onto the bus. The 74374 is a TTL octal D-type flip-flop device.

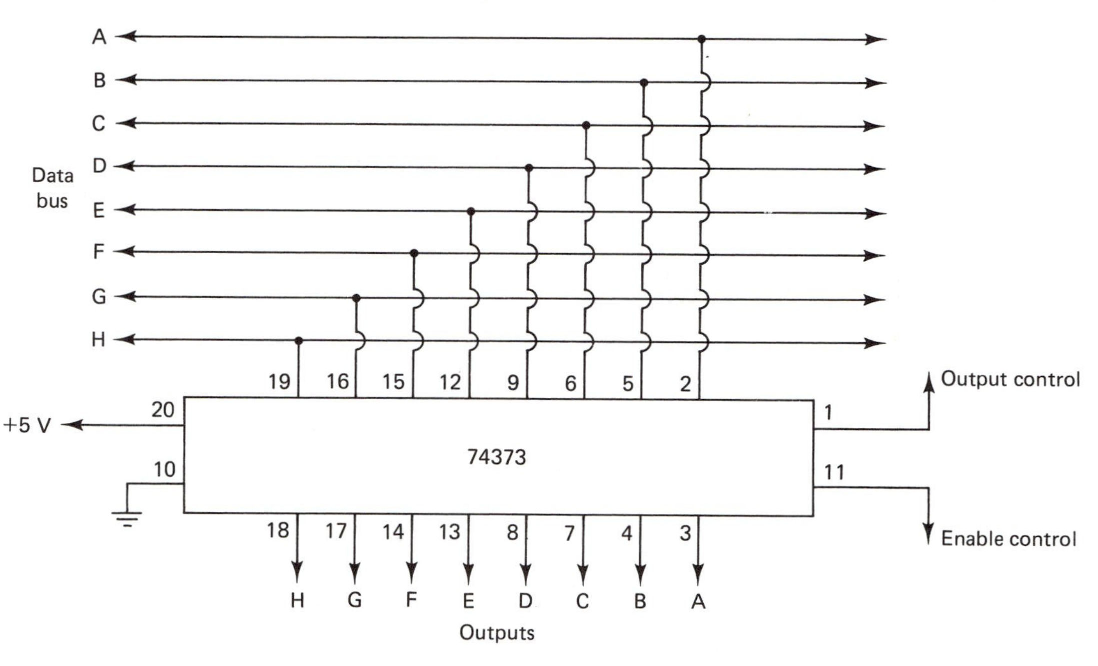

Figure 36 Bus register

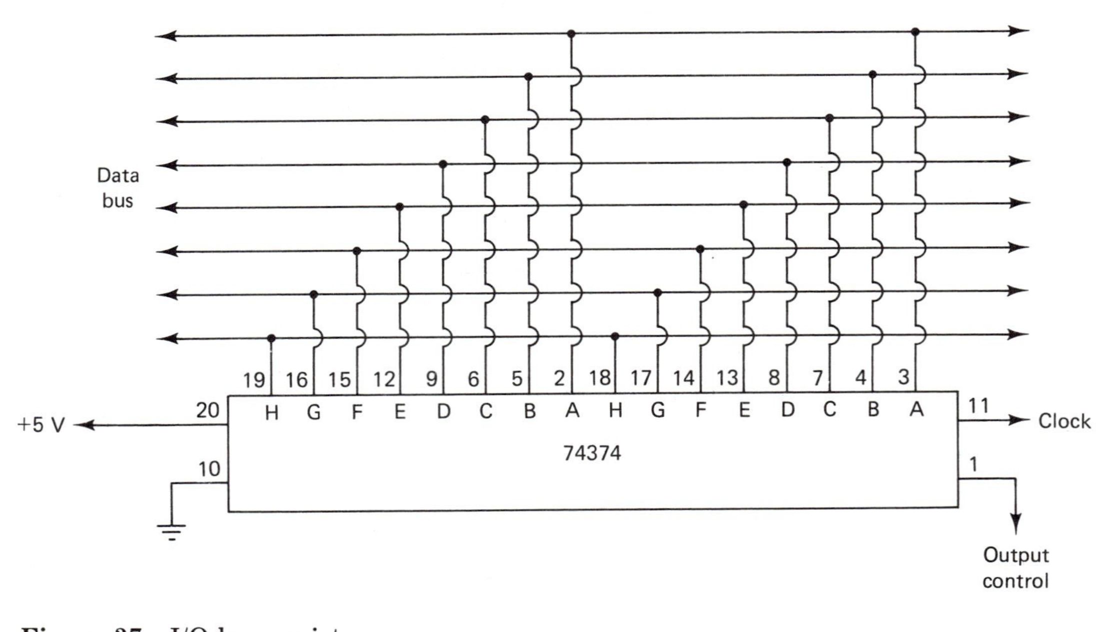

Figure 37 I/O bus register

Bus selector

This circuit, built around a 74157 TTL quad data selector device, allows for selection between two four-bit buses and reading the signals on them. The choice of which bus signals appear at the outputs is made by applying a high or low signal to the bus select input. A low signal selects Bus 1, a high signal Bus 2. The circuit, shown in Figure 38, does not affect any of the signals on either bus, nor does it affect the operation of either bus network.

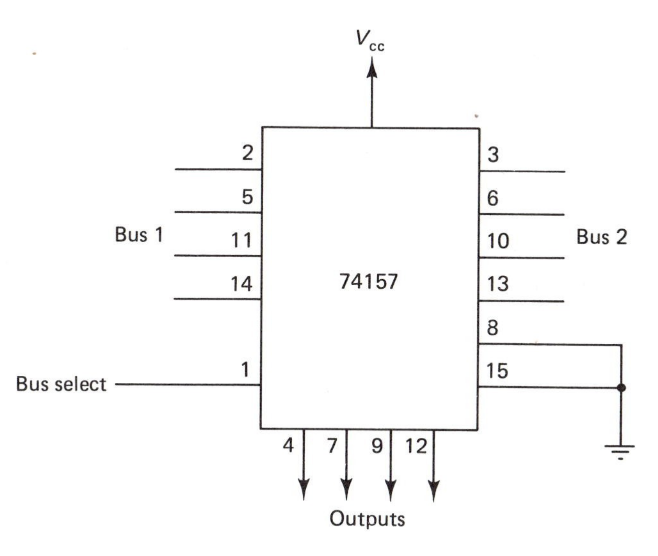

Figure 38 Bus selector

Bus transceiver

A bus transceiver is a circuit which allows two different data buses to "talk" to each other. The circuit in Figure 39 uses a 74245 TTL octal bus transceiver device to provide communication between two eight-bit buses. The operation is controlled by signals to the enable and bus direction inputs. When the signal to the enable input is low, communication between the buses is possible. When it is high, the two buses cannot communicate. The bus direction input controls the direction in which communication takes place when the enable input is low. If the bus direction input signal is low, data flows from Bus 1 to Bus 2. When it is high, data travels from Bus 2 to Bus 1.

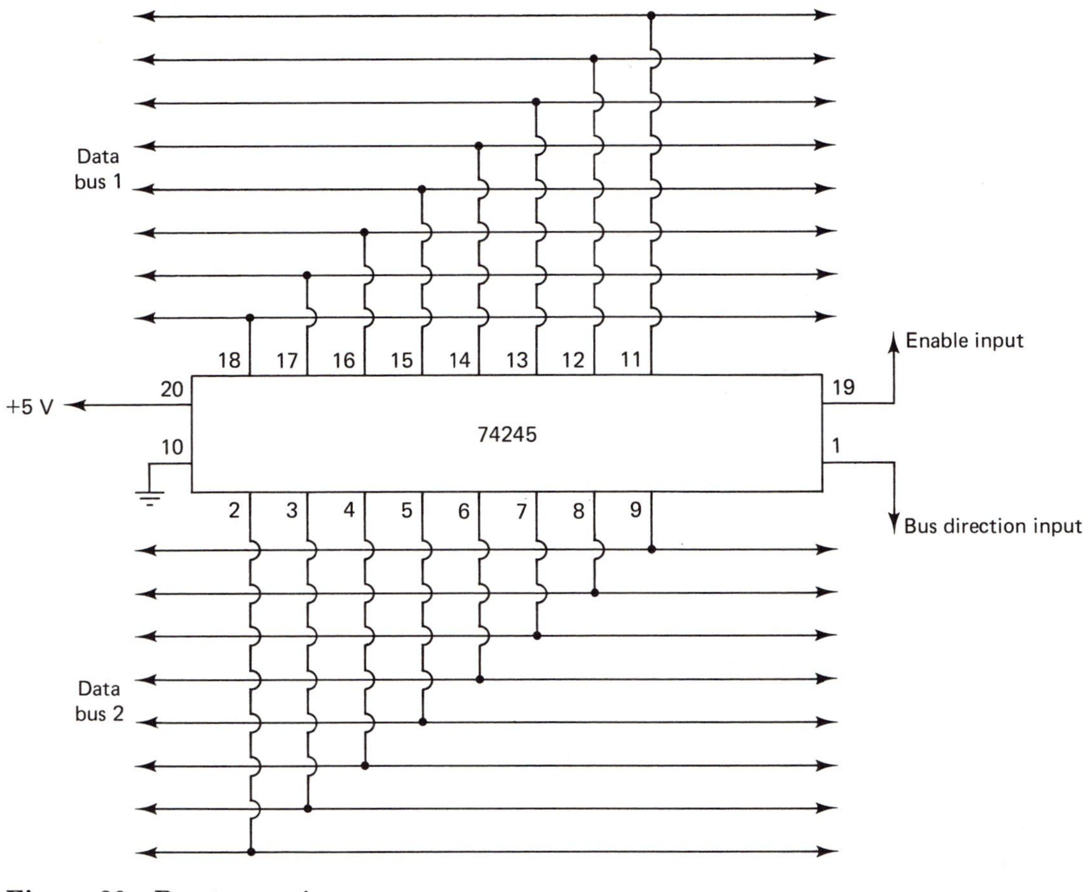

Figure 39 Bus transceiver

Bus transfer

The circuit in Figure 40 permits data to be read from and written to a four-bit data bus. The heart of the circuit is a 74244 TTL non-inverting octal buffer. If the read control input (pin 19) is low, data is transferred from the bus and appears at the "from bus" outputs. If the write control input (pin 1) is low, data present at the "to bus" inputs is transferred to the bus. The other control input should be kept high: if both are high, the device is isolated from the bus.

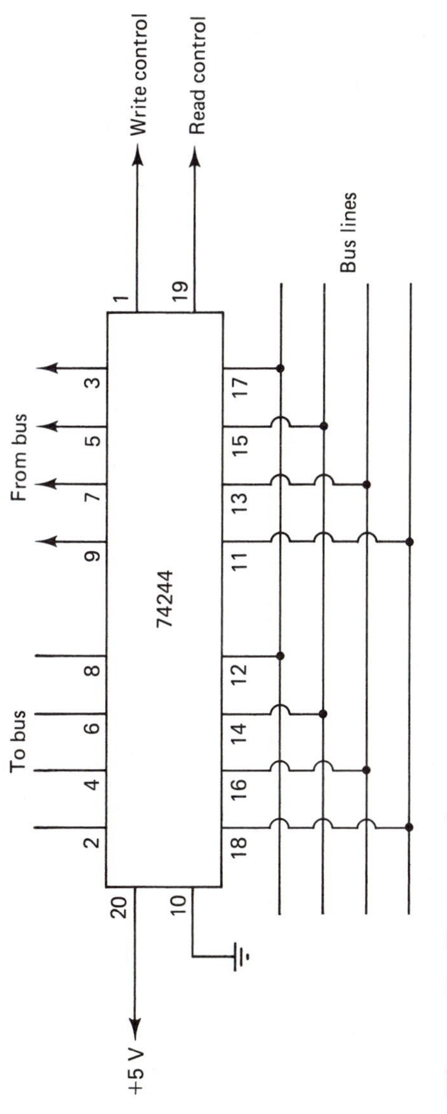

Figure 40 Bus transfer

C

Clippers, active

A clipper circuit "shaves off" the peaks of an input signal. For example, a sine wave input would produce an output with flat peaks and gently sloping sides, much like a misshapen square wave. Clippers may be constructed using strictly passive components. However, they may also be built using op amps, and this is a preferable approach when dealing with weak input signals since such circuits can clip and amplify input signals.

Figure 41 shows an inverting active clipper. The clipped output

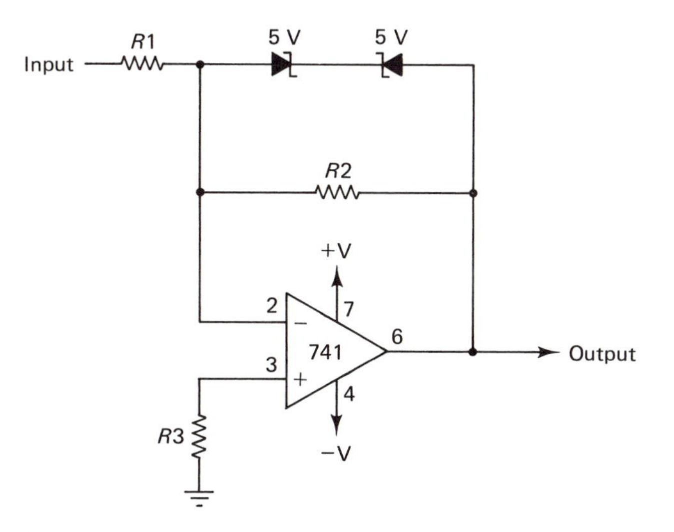

Figure 41 Inverting active clipper

is of opposite polarity to the input waveform. The gain of this circuit is determined by resistors $R1$ and $R2$ according to the formula

$$\text{Gain} = -\frac{R2}{R1}$$

$R3$ is added to the circuit to stabilize the amplifier and prevent oscillation. Its value is given by

$$R3 = \frac{R1R2}{R1 + R2}$$

Figure 42 shows a noninverting clipper. The gain of this circuit is given by

$$\text{Gain} = 1 + \frac{R2}{R1}$$

All diodes in Figures 41 and 42 are Zener types rated at 5 V.

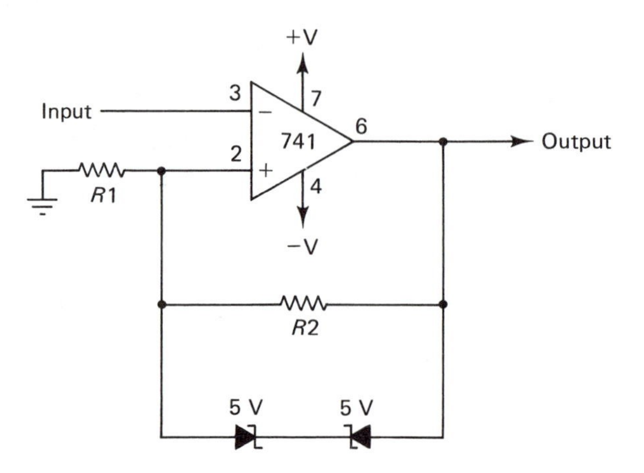

Figure 42 Noninverting active clipper

Clippers, passive

If an adequate input signal is available, a passive clipper is a simpler and cheaper solution than an active clipper. Figure 43 shows two passive clippers. The positive peak clipper produces an output which is a clipped positive waveform, while the symmetrical clipper produces an output of clipped positive and negative peaks.

Each circuit uses Zener diodes. At a minimum, they should be rated at the maximum input voltage expected, and it is good practice to comfortably exceed the maximum expected input voltage in selecting them. The value of the resistor R is given by the formula

$$R = \frac{\text{input voltage} - \text{Zener voltage}}{\text{current through } R}$$

The wattage rating of R should also comfortably exceed that expected, as calculated using Ohm's law.

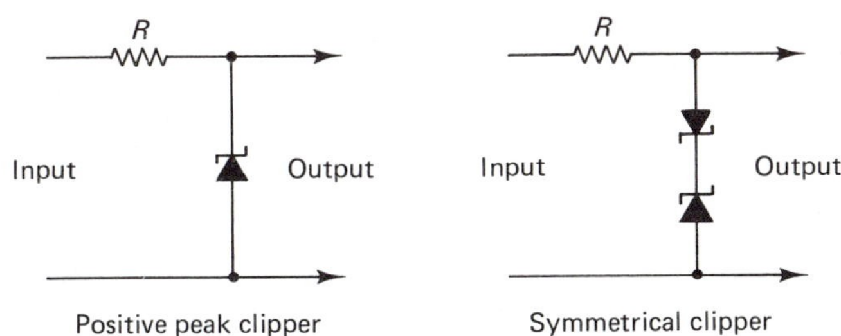

Figure 43 Passive clippers

Clock signal generators

Clock signal generators are circuits which produce a square wave output which is used to control the operation of various sequential digital circuits. Clock signal generators come in numerous configurations.

Figure 44 shows a clock generator for use with CMOS circuits. The device is made from two sections of a 4049 inverting hex buffer IC, and the clock frequency is controlled by the values of *R* and *C*. The formula for finding the clock frequency is

$$\text{Clock frequency} = \frac{1}{1.4RC}$$

The value of *C* should limited to between 0.01 to 10 μF.

A clock generator suitable for TTL circuits is shown in Figure 45. It is built using all four gates of a 7400 NAND gate IC and is capable of operation from 100 kHz to 3 MHz. The output frequency is determined by crystal X, which is cut to the desired output frequency. The output frequency is "tweaked" by the 110 pF variable

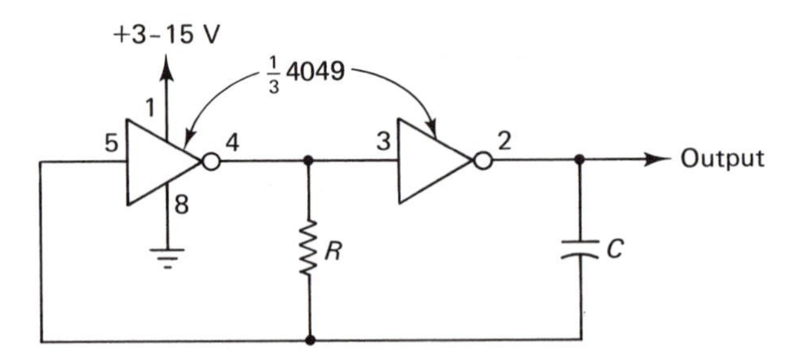

Figure 44 CMOS clock signal generator

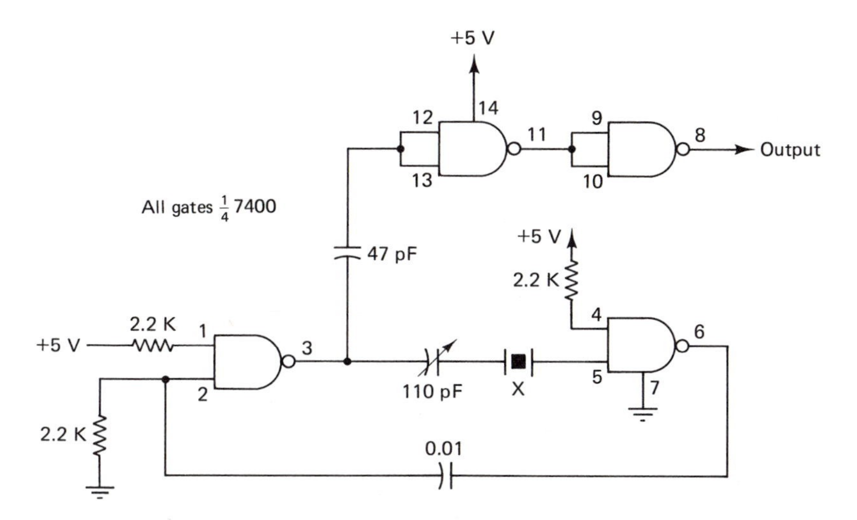

Figure 45 100 kHz–3 MHz TTL clock generator

capacitor for maximum accuracy. A crystal socket can be used to allow changing crystals for different output frequencies.

Parts placement and layout can be crucial in both clock signal generators shown. Also, leads should be kept as short as possible.

Comparator

A comparator can be thought of as an operational amplifier which has been configured to work as an on-off switch. The device works as an op amp operating in the *open-loop* mode; this means that even the smallest input signal will cause maximum output from the circuit. Thus, any input produces a "full blast" output. The most common application of a comparator is to detect differences between an input voltage and a reference voltage: if the voltages differ, an output is produced. While ordinary op amps can be used as comparators, it is more common to use op amps specifically designed for that purpose, such as the 339, which contains four comparators.

Figure 46 shows noninverting and inverting comparators with hysteresis. If the input voltage differs from the reference voltage, the output of these comparators immediately goes to maximum. The output polarity of the noninverting comparator is the same as that of the input, while it is the opposite in the case of the inverting comparator. Hysteresis provides a feedback loop to prevent oscillations, which can degrade circuit performance.

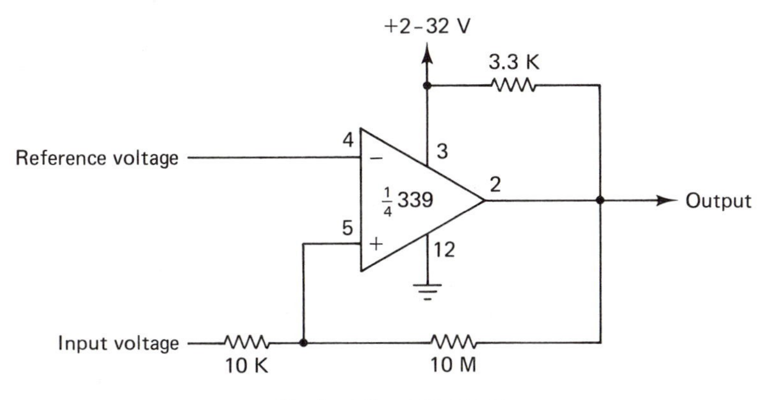

Noninverting comparator

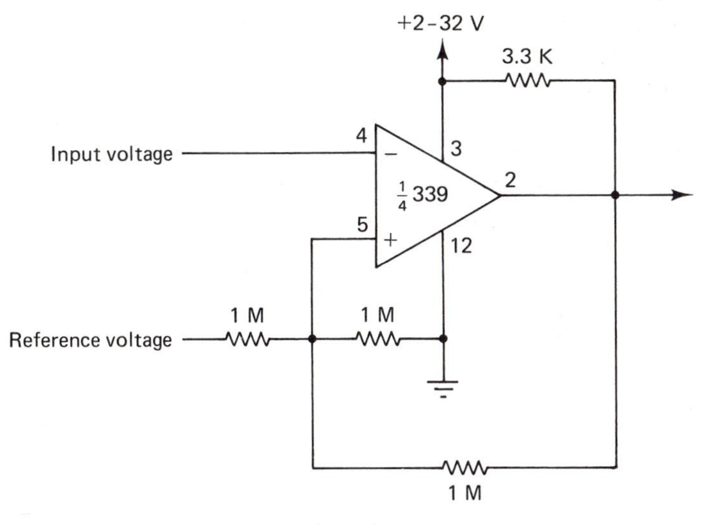

Inverting comparator

Figure 46 Noninverting and inverting comparators with hysteresis

Comparator, window

Conventional comparators produce an output whenever the input voltage varies from a single reference voltage. The circuit in Figure 47 is a variation known as a *window* comparator; two reference voltages are provided to form a "window" over which the input voltage can fall. The circuit normally has an "on" output. If the input voltage falls into the range between the high and low reference voltages, the output is turned off. The output can be fed to an inverter if a reversed polarity output is needed. The input and reference voltages must not exceed the supply voltage.

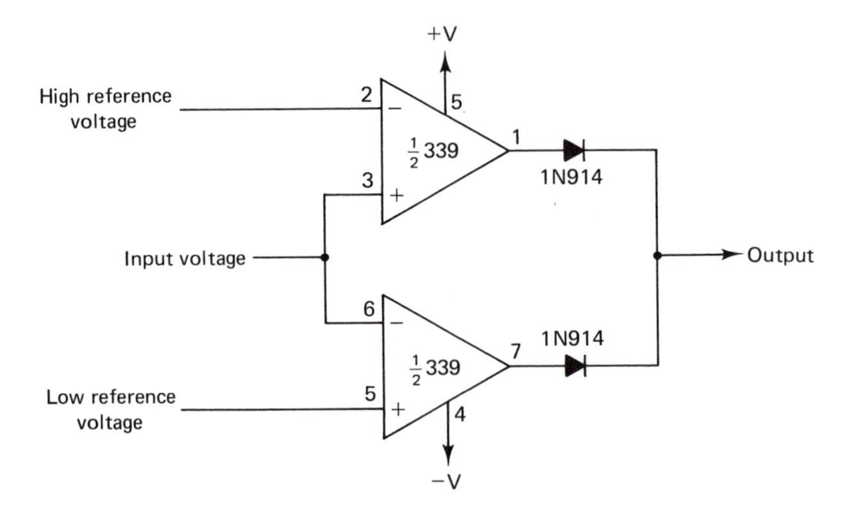

Figure 47 Window comparator

Comparator, word equality

Comparator circuits designed for analog signals, such as the preceding circuit, are not suitable for many digital applications. Digital data is frequently grouped into "words" of four, eight, or more bits, and operations often involve comparisons between different words, such as whether two words are equal or not. Some ICs, such as the 74LS688, have been developed to simplify the circuitry required to perform these comparisons. The 74LS688, an LSTTL device, can accept two eight-bit binary or BCD words as inputs. Its output, at pin 19, will remain high if the words are unequal or if there are not two words present at the inputs. If the words at the inputs are equal, the output goes low. Pin 1 of the device is a cascade input, shown tied to ground in Figure 48. Several 74LS688 devices can be used to compare words of more than eight bits by connecting the output from pin 19 to the cascade input of a following 74LS688. Additional bits making up words in excess of eight bits are placed at the inputs of the following device.

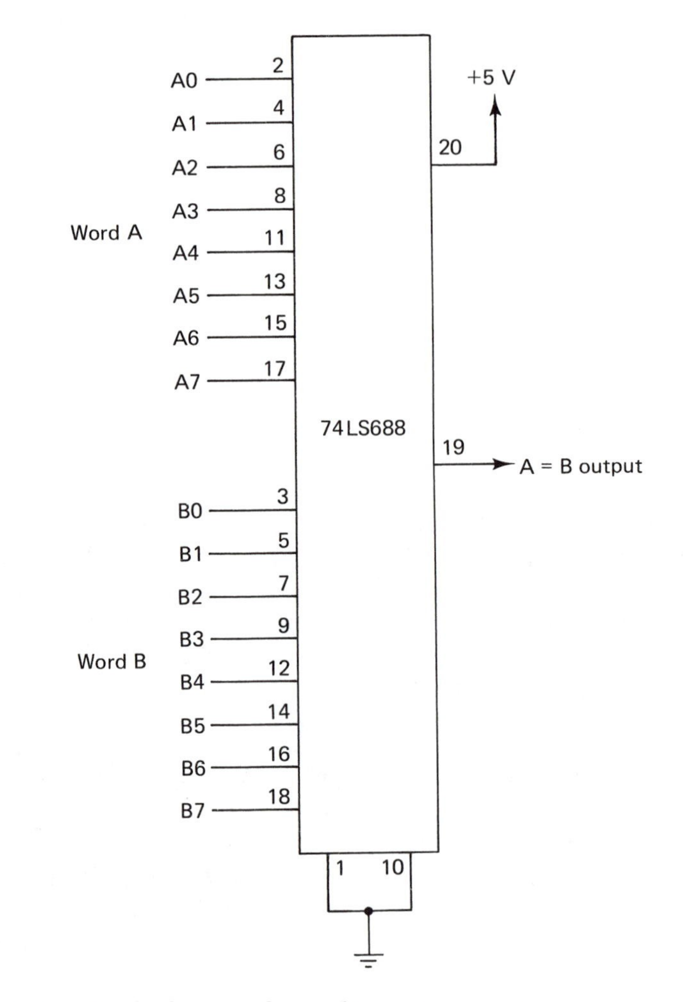

Figure 48 Eight-bit word equality comparator

Comparator, word magnitude

Often, we need to know more about the relationship between two data words than just whether they are equal or not; we may also want to know whether, if they are not equal, one is greater or less than the other. A TTL device intended for such purposes is the 74LS85 four-bit word comparator IC. Figure 49 shows this IC, which accepts two four-bit words at its inputs and produces one of three outputs, $A > B$, $A = B$, and $A < B$. Depending upon the status of the four-bit words at the inputs, the appropriate one of these outputs will be high and the other two will be low.

It is possible to cascade two or more of these devices to compare words of more than four bits. Figure 50 shows how this can be done to compare two eight-bit words. The three outputs of the 74LS85 are fed to the corresponding cascading input of another 74LS85. Pin 4 is the $A > B$ input, pin 3 is the $A = B$ input, and pin 2 is the $B < A$ input. When cascaded as shown, the words are "split." The least significant bits (LSBs) of each word are applied to the inputs of the first device, while the most significant bits (MSBs) are applied to the last devices. By combining the 74LS85 devices shown here, two data words of considerable length can be compared.

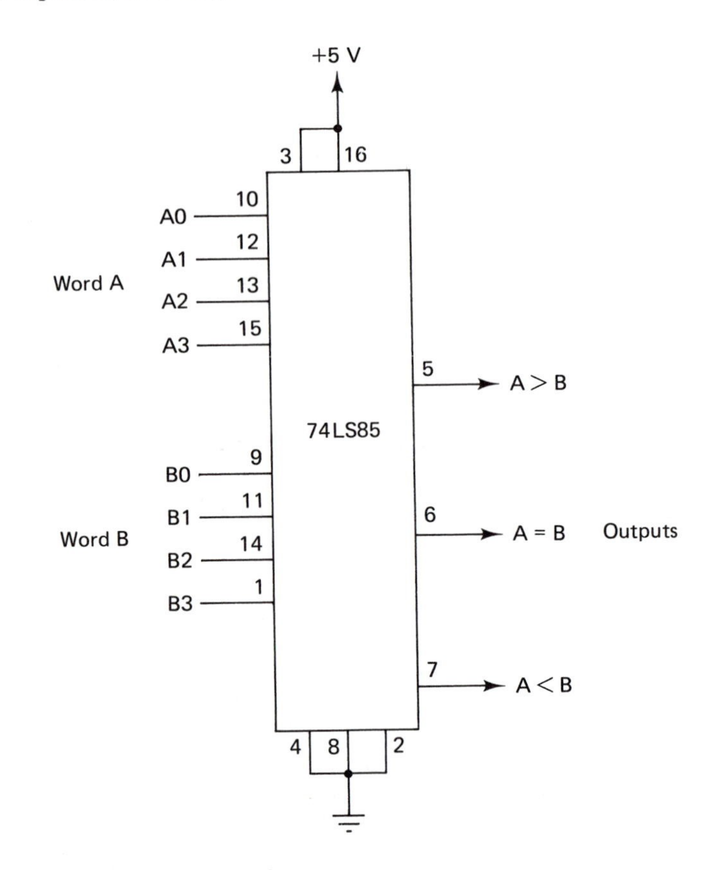

Figure 49 Four-bit word comparator

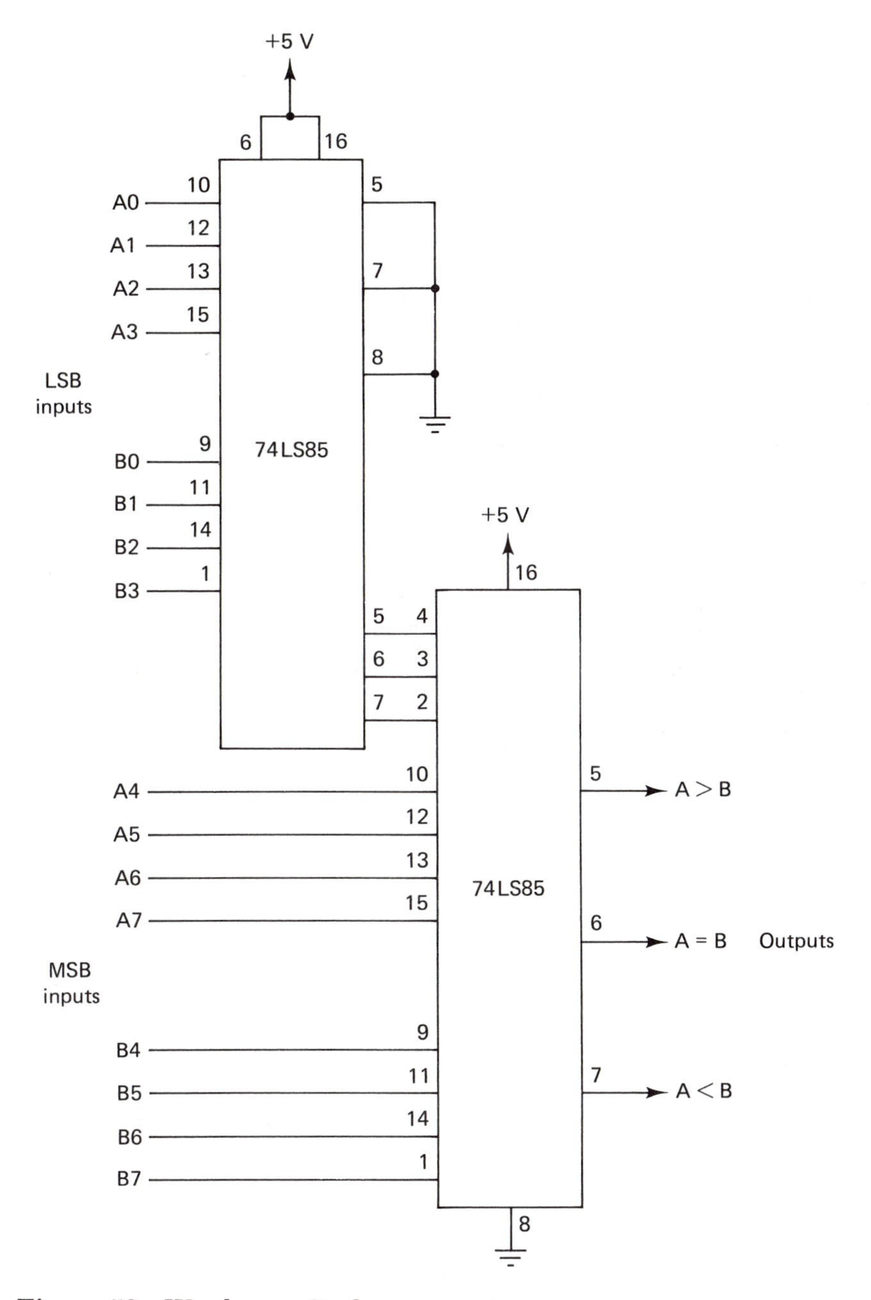

Figure 50 Word magnitude comparator

Compressor, audio

An audio compressor is often used as part of a complete *speech processing* system. The compressor circuit takes an audio signal and "compresses" it so that the ratio between the peak and the average level of the signal is reduced. When such a signal is used to modulate a radiotelephone transmitter, the result is a higher average modulation level. When the signal is received, another circuit known as an expandor restores the proper ratio of peak to average power.

The circuit in Figure 51 uses one-half of an NE571 AGC device.

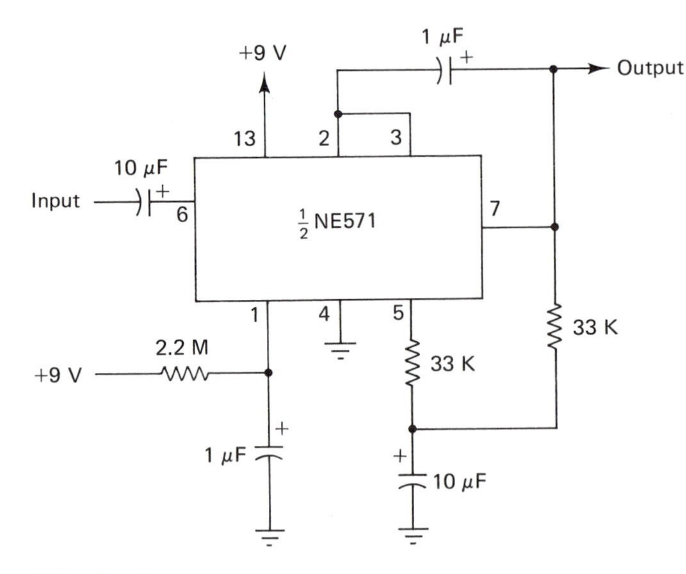

Figure 51 Audio compressor

The AGC control voltage is derived from the output signal, forming a closed loop. The output follows the average, rather than instantaneous, variations in the audio input signal. The input signal frequencies can be any in the normal audio range, although best performance will be with normal human voice frequencies (3,000 Hz and lower).

Converter, analog-to-digital

Analog-to-digital conversion (ADC) can be accomplished by a variety of methods, but the trend has been toward ADC ICs which hold all the necessary circuitry in a single package. Numerous ADC ICs have been developed for various applications, but the most commonly used may well be the ADC0802, which takes an analog input and produces an eight-bit TTL output. The outputs are normally connected to an eight-bit microprocessor bus, and three control signals are received from the microprocessor. However, for experimental purposes, other TTL-level signals (such as from TTL devices or mechanical switches) may be substituted. Figure 52 shows a typical configuration for this circuit. The reference voltage is 2.5 V and can be supplied from an LM336 voltage reference device or its equivalent. The analog input voltage can range from 0 to 5 V, and the conversion accuracy is typically ± 0.5, the least significant bit (LSB) of the eight-bit word produced. Note that digital and analog ground points are used.

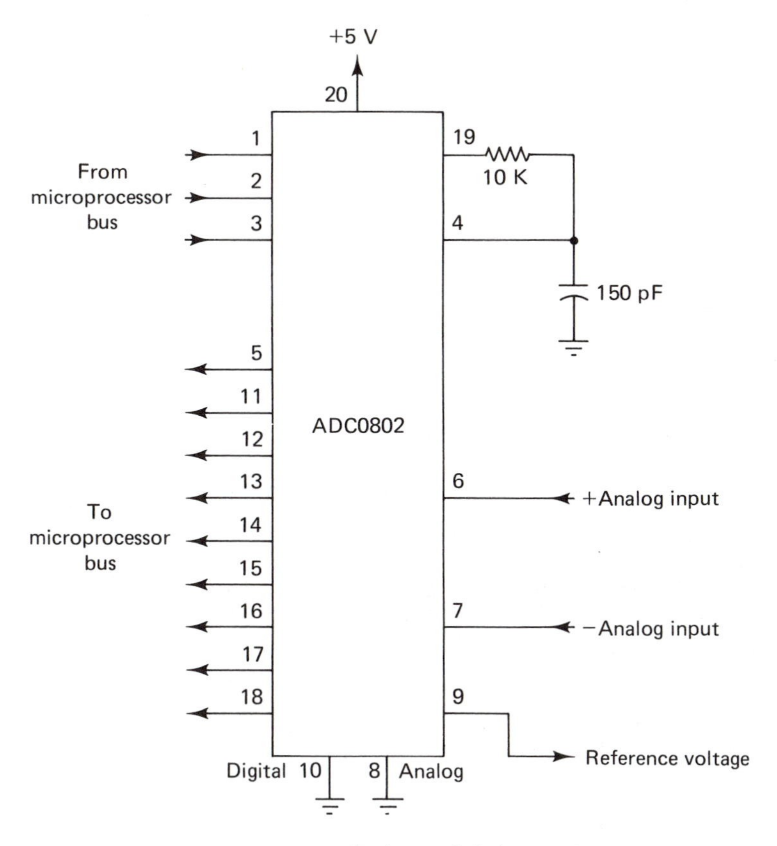

Figure 52 Analog-to-digital converter

Converter, DC-to-DC

Converting one DC voltage to another can present problems, especially if both polarities are needed. One approach that is becoming increasingly popular is to use a pulse width modulator (PWM) IC. The circuit shown in Figure 53 will take a dual-polarity 12 V source and produce a dual-polarity 5 V output. Maximum output current is typically 100 mA. Several precision (1 percent) resistors are used, because substitution of resistors with greater tolerance will result in an output that can vary considerably from the intended 5 V. Regulation is typically 0.2 percent or greater.

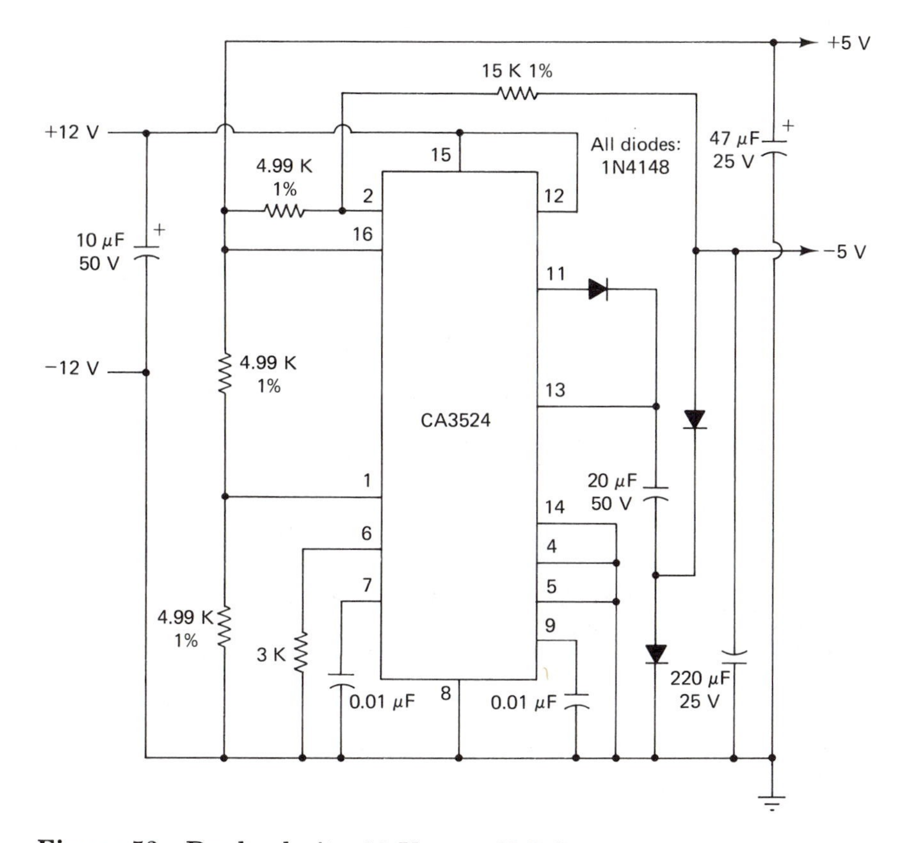

Figure 53 Dual-polarity 12 V-to-5 V DC converter

Converter, digital-to-analog

As in analog-to-digital conversion, digital-to-analog conversion (DAC) is increasingly being done by ICs specifically designed for that purpose. A very popular DAC IC is the DAC0800, which takes an eight-bit digital input and produces an analog output which ranges from 0 to ± 20 V, peak to peak, depending on the digital input. Figure 54 shows a typical DAC circuit using the DAC0800, which offers an accuracy of plus or minus the LSB of the digital word input. The digital inputs may be TTL or CMOS, and the time it takes to perform the conversion (the *settling* time) is 100 nS or less. The dual-polarity supply voltages may range from 4.5 to 18 V.

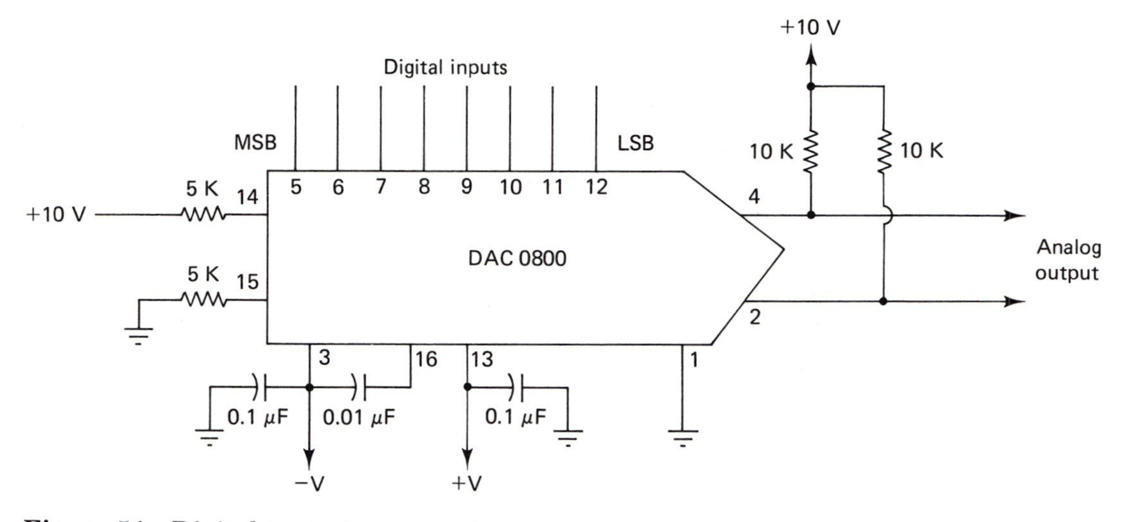

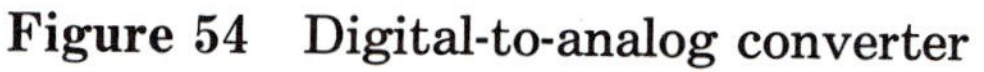
Figure 54 Digital-to-analog converter

Converter, frequency-to-voltage

A frequency-to-voltage converter produces an output voltage which varies in accordance with the frequency of an input signal. Several ICs intended for this application have been developed. One is the LM331, which can be used for either frequency-to-voltage or voltage-to-frequency conversion. Figure 55 shows one circuit using the LM331 which gives an output of 1 V for every 1 kHz of frequency at the input. The circuit is surprisingly linear, with the 5 K potentiometer adjusted to provide the most accurate operation. The circuit accuracy rises as the supply voltage increases; thus, a 15 V supply voltage provides a more accurate output than a 9 V source.

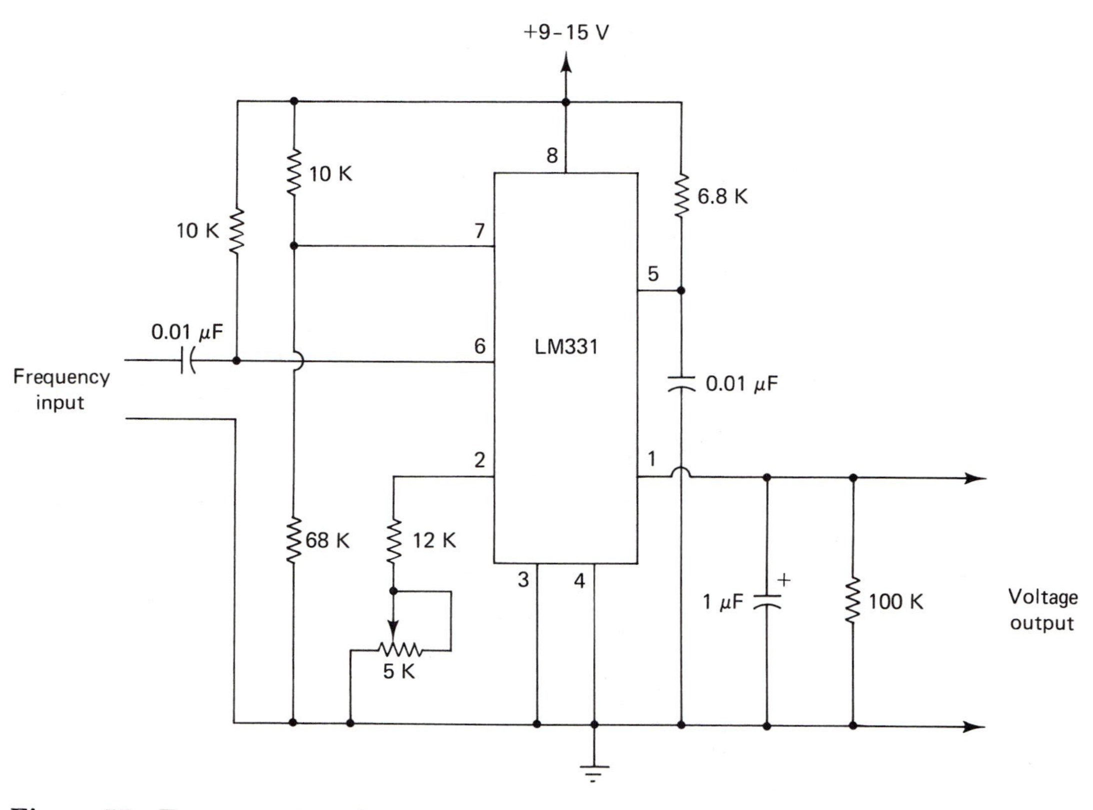

Figure 55 Frequency-to-voltage converter

Converter, parallel-to-serial

A parallel-to-serial converter accepts data on parallel input lines and converts it into a data stream, one bit following another, sent over one serial line. The easiest way to accomplish this is through the use of a parallel in-serial-out shift register IC, such as the 4021 CMOS eight-stage shift register. Figure 56 shows the 4021 configured to accept data from an eight-bit data bus and produce a serial output. A switch is included to allow a choice between loading and sending data. When the switch is set to load, pin 9 becomes high and data at the parallel inputs is loaded into the register. When the switch is set to send, pin 9 is connected to ground and the contents of the shift register "empty" into the serial output. After the con-

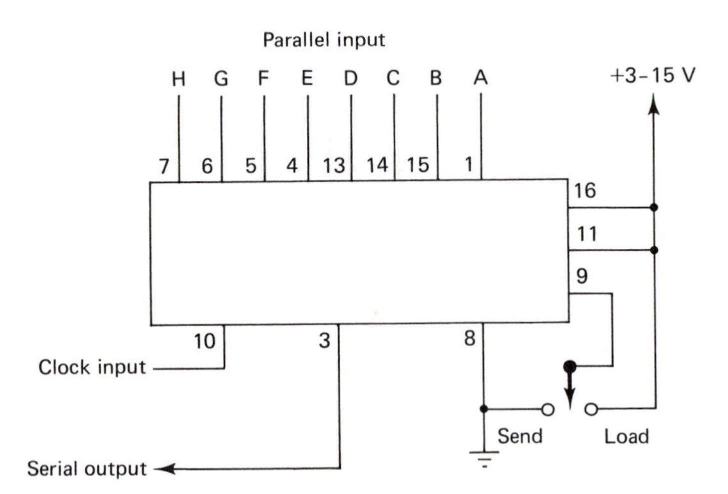

Figure 56 Parallel-to-serial conversion

tents of the shift register have been transmitted, a stream of high bits is transmitted at the clock signal rate until a new parallel input is loaded and transmitted. Pin 9, the load control pin, overrides the clock signal. When pin 9 is high, the parallel inputs are loaded into the shift register regardless of what the clock signal is; when pin 9 is low, parallel inputs are loaded with each clock pulse. For normal (unswitched) operation, pin 9 should be kept low.

Converter, serial-to-parallel

A serial-to-parallel converter performs the opposite function of the parallel-to-serial converter, taking a stream of input data from a single data line and dividing it into a given bit format (such as four or eight bits) for transmission along a data bus in parallel mode.

Figure 57 shows a circuit capable of producing an eight-bit parallel output from a serial input. This circuit uses a 7490 divider IC,

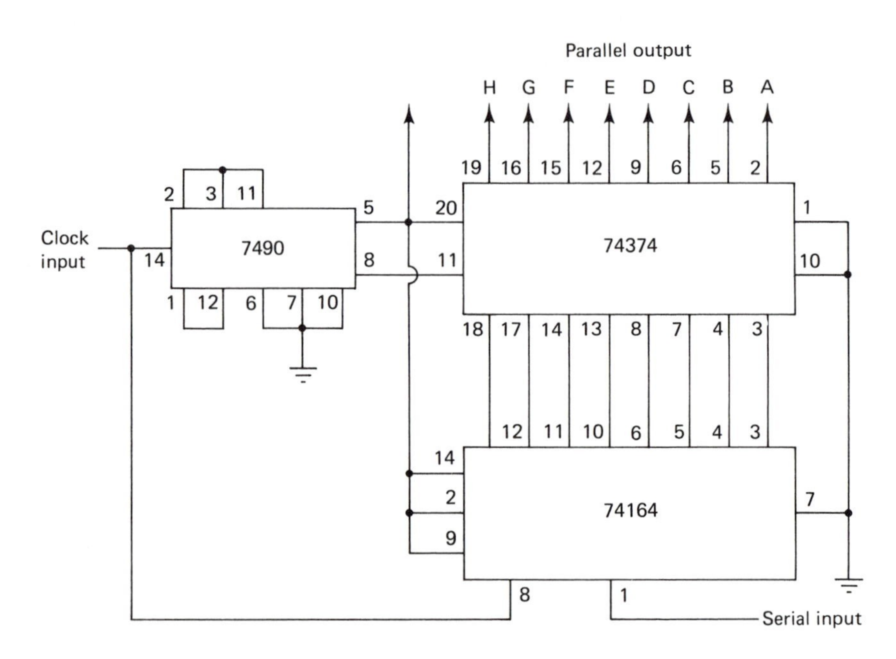

Figure 57 Serial-to-parallel (eight-bit) converter

a 74164 eight-bit shift register, and a 74374 octal D-type flip-flop device. The 7490 is configured as a divide-by-eight counter and thus divides the clock input by eight. *The serial input rate to the 74164 is the same as the clock signal rate.* The output of the 74164 is loaded into the 74374 at the eight-bit intervals: each time a full eight-bit output is available from the 74164, the entire eight bits are loaded into the 74374 and are then transmitted to the bus from the 74374 outputs.

Converter, triangle-to-sine-wave

A triangle-to-sine-wave converter takes a triangular or "sawtooth" waveform and produces a sine wave output. While it is simple to implement a circuit to perform this function to a limited degree, a high-performance version (one that offers a relatively undistorted sine wave output) is more difficult to achieve.

Figure 58 shows a circuit using both halves of a CA3280 variable operational amplifier IC. The output has a total harmonic distortion of 0.37% or less for an input signal of 170 mV peak to peak. Two 100 K resistors are connected between differential amplifier emitters of the op amps and the positive supply voltage to reduce current flow through the differential amplifiers, allowing the amplifiers to fully cut off during peak input signal excursions.

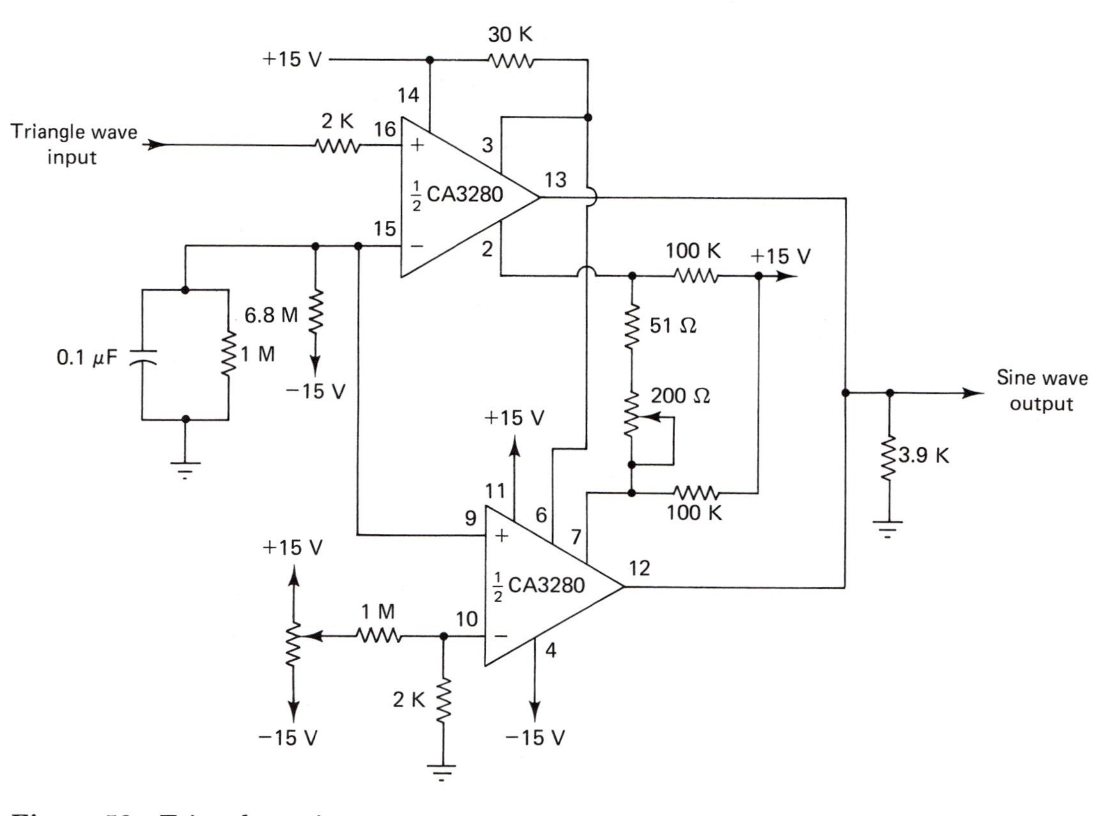

Figure 58 Triangle-to-sine-wave converter

Converter, voltage-to-frequency

A voltage-to-frequency converter produces an output signal whose frequency varies with the voltage applied to its input. Two popular approaches to implementing these circuits are to use phase-locked loop (PLL) ICs, or ICs specifically designed to be used as voltage-to-frequency converters.

Figure 59 shows a voltage-to-frequency converter built around the 4046 PLL IC. Adjusting the 500 K potentiometer will vary the output frequency from approximately 0 Hz to 18.5 kHz. For best results, *V*dd should be 15 V. The 500 K potentiometer may be eliminated if another method of controlling the input voltage is used. While this circuit is simple, its output will usually be highly nonlinear.

Better performance can usually be obtained by using a device designed specifically as a voltage-to-frequency converter. One popular such IC is the XR-4151. Figure 60 shows a simple circuit using

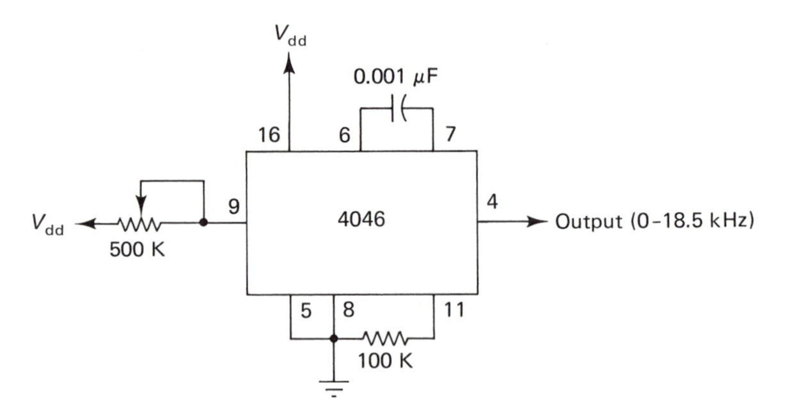

Figure 59 Basic voltage-to-frequency converter

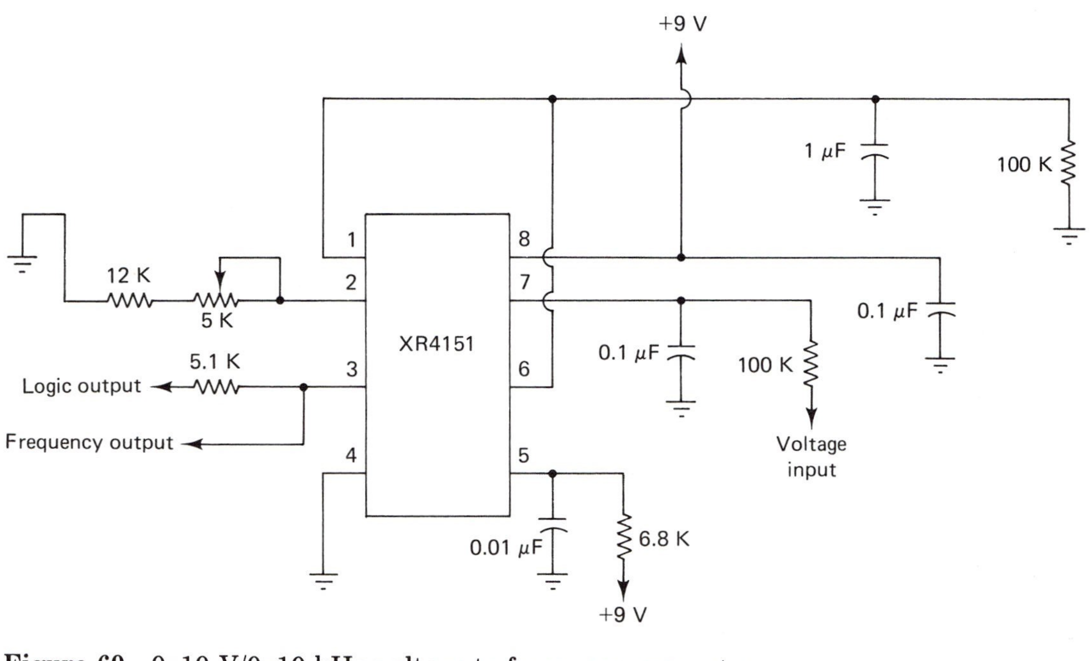

Figure 60 0–10 V/0–10 kHz voltage-to-frequency converter

this device which produces an output frequency of 0 to 10 kHz for an input voltage of 0 to 10 V. The maximum output frequency can be set by the 5 K potentiometer. This circuit also has a "logic output": an internal voltage comparator in the XR-4151 compares a positive input voltage applied at pin 7 to the voltage at pin 6; if the input voltage is higher, the "one shot" fires. The linearity of the circuit is typically 1 percent.

Counters/dividers

A counter/divider is so termed because it can be described as either a counter of a stream of input pulses or a divider of that stream. Counters/dividers produce a single output pulse for a given number of input pulses. For example, a divide-by-10 counter/divider produces one output pulse for every 10 input pulses. Viewed from one perspective, such a circuit divides the input pulses; but viewed from another perspective, it can be thought of as counting the input pulses and signaling (via a single output pulse) when it has reached 10. Regardless of whether a circuit counts or divides, it is commonly referred to as a "divide by *N*" counter.

Several divider/counter TTL and CMOS ICs have been developed and are in wide use. The output of most counter/divider ICs can serve as the input of another counter/divider; this "chaining" or cascading allows higher counting or division than is possible with a single device and is indeed quite common.

A very popular TTL device is the 7490, which can be configured for several division or counting ratios. Figure 61 shows the most popular configurations; the highest ratio possible with this device is 10. *The 7490 may not be cascaded.*

Another popular TTL device is the 7492, which is a divide-by-12 IC. Figure 62 shows the 7492 configured first as a divide-by-12, and then as a divide-by-120, counter. The divide-by-12 counter shows the 7492 configured in what might be termed its "normal" mode of operation. In the divide-by-120 application, the first 7492 is configured as a divide-by-10 counter while the second is used as a divide-by-12 counter. The output of the first 7492 is one pulse for each 10 input pulses; this output is then used as the input to the second 7492. The second 7492 produces one output pulse for every 12 input

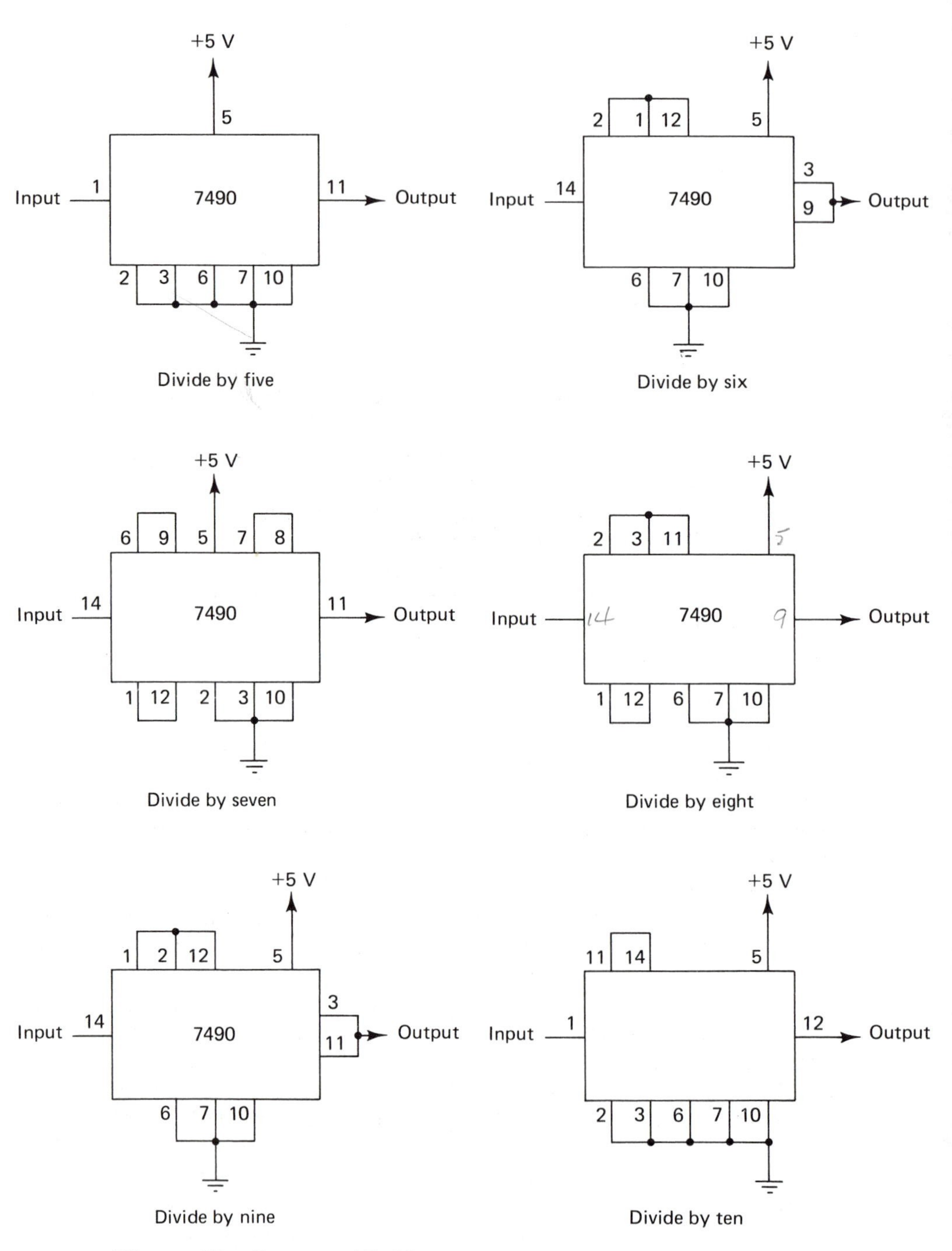

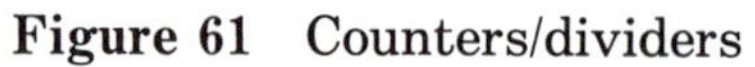
Figure 61 Counters/dividers

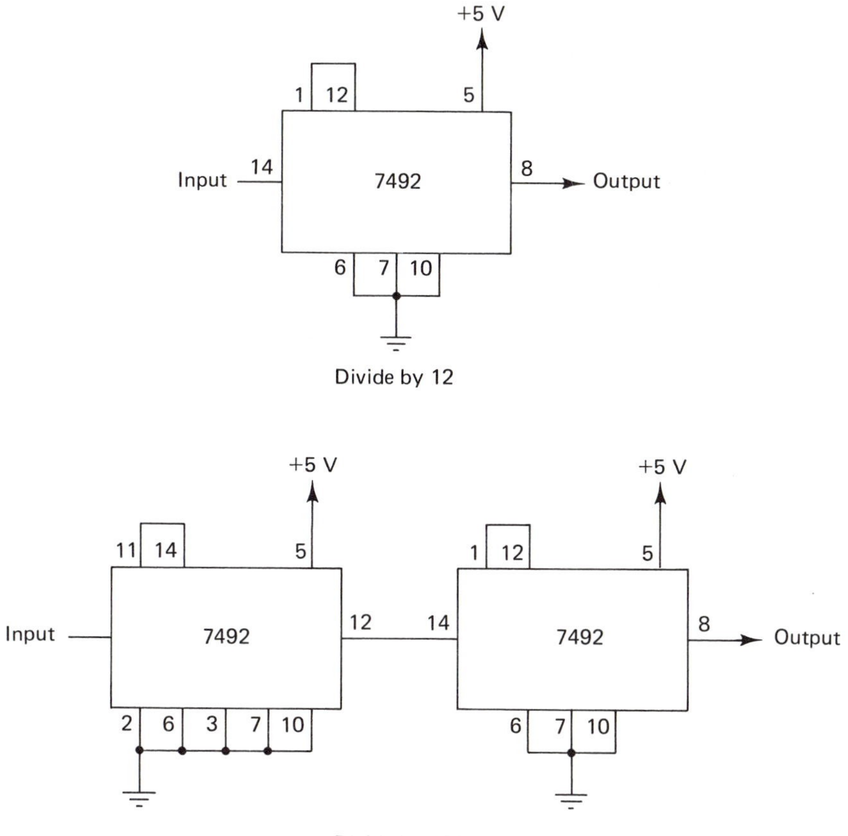

Figure 62 Counters/dividers

pulses. The result is one output pulse from the complete circuit for every 120 input pulses.

CMOS counter/divider ICs are also available. Two of the most common are the 4017 and 4018. The 4017 is a divide-by-10 device with a one-of-10 decoded output, meaning that one of the outputs sequentially goes high while the others remain low to indicate the count. The 4018 is capable of division by any ratio between two and 10. Figure 63 shows typical applications of these devices.

The 4017 is readily cascaded, and the decoded outputs make possible a counter with outputs available from 0 to 99. When the switch is set to "RUN," the circuit will count the input clock signal and produce a high output at the pins indicated depending upon the status of the count. When the switch is set to "RST," the counter

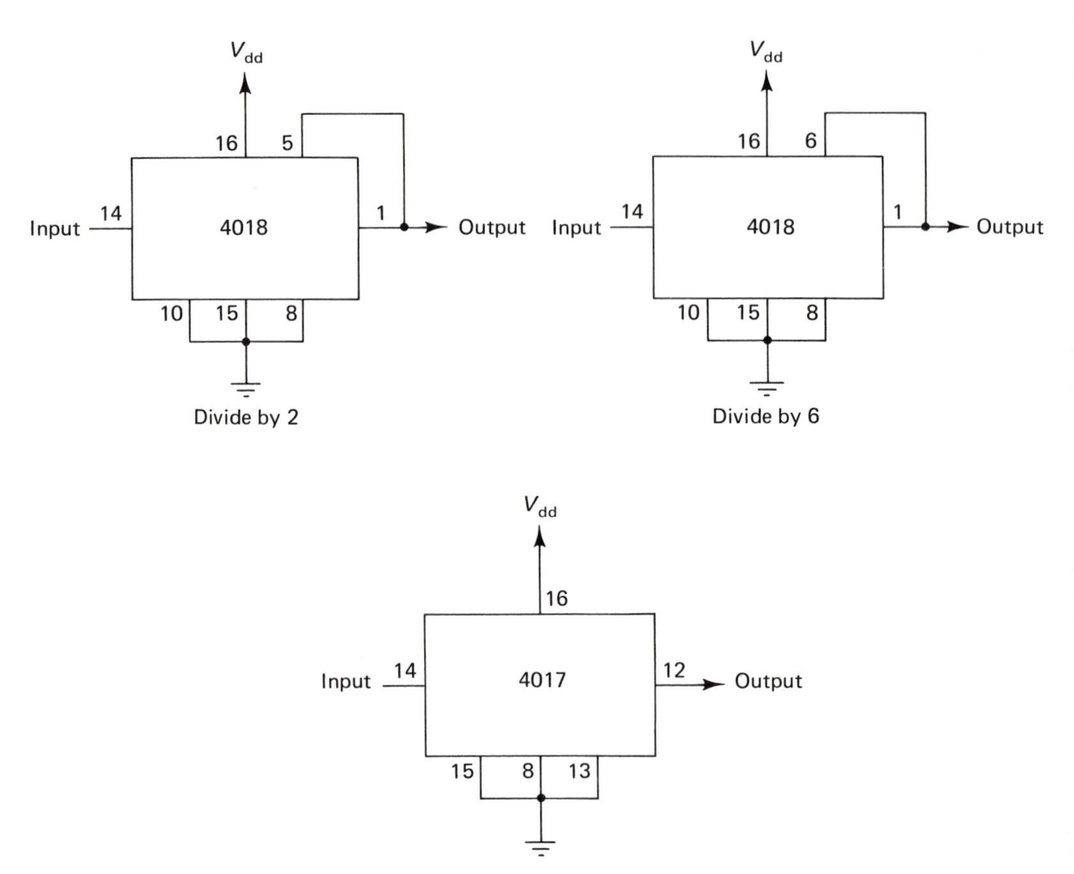

Figure 63 Counters/dividers

is cleared and the count begins over at zero. The count also begins over at zero when 99 is exceeded.

Three 4017 devices can be cascaded to produce a counter/divider providing outputs which are equal to the input signal divided by 10, 100, and 1,000. Figure 65 shows this arrangement. For simplicity, the decoded outputs are not shown in the diagram, but they are available if needed.

Another interesting counter/divider IC is the 4024, a so-called seven-stage counter capable of simultaneously dividing an input signal by 2, 4, 8, 16, 32, 64, and 128. Figure 66 shows how the 4024 can be used to produce such outputs, which can be used as the inputs for other CMOS counter/divider devices.

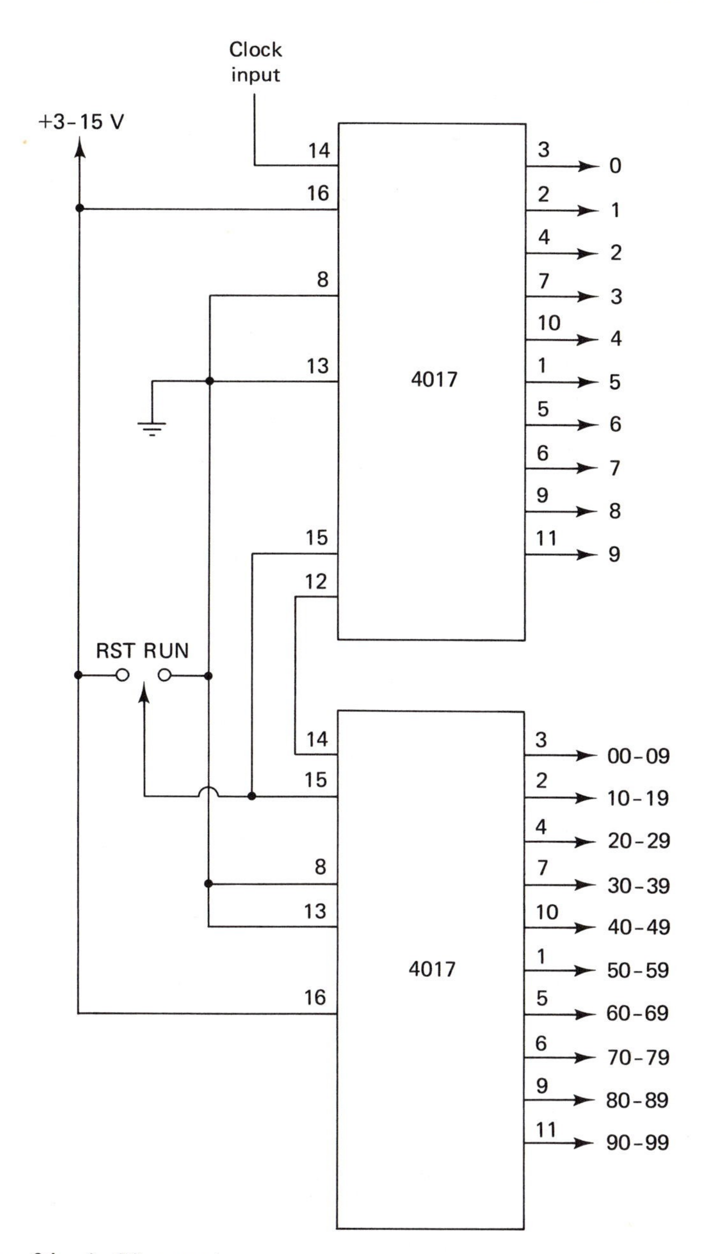

Figure 64 0–99 counter

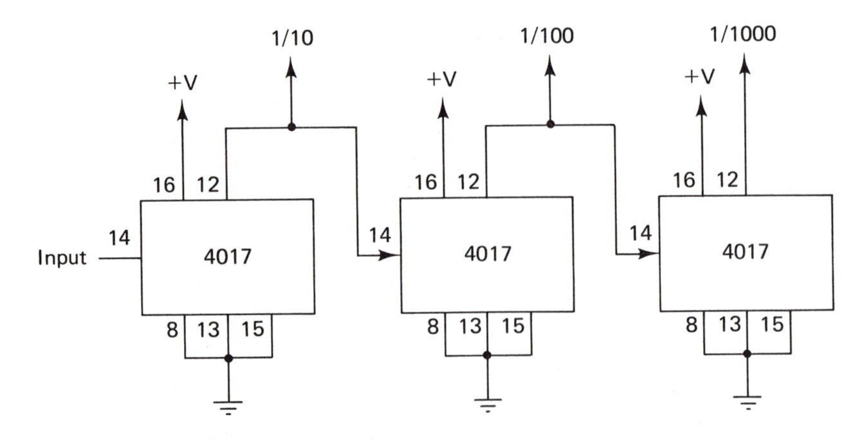

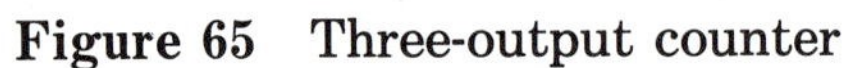

Figure 65 Three-output counter

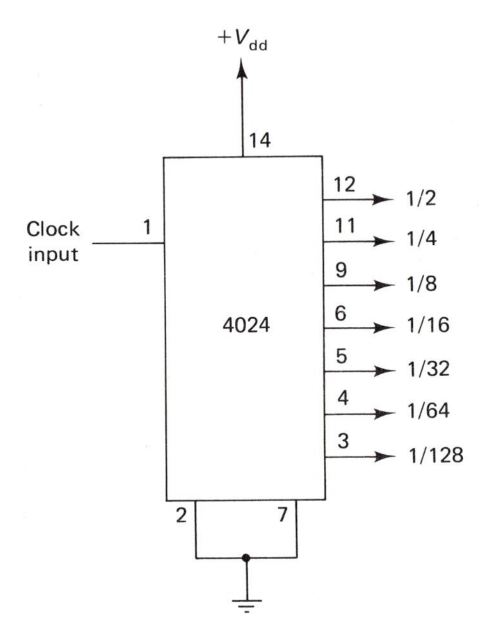

Figure 66 Divide-by-128 counter

Current source, voltage-controlled

It is easy to make an op amp furnish a certain level of output current for a given input voltage, as long as the output current and input voltage remain at a fixed, predetermined level. But what happens when we need to vary the output current over a range of possible values? One approach to solving this problem is shown in the circuit of Figure 67, which uses a TL081 JFET input op amp. Note that five resistors labeled $R1$ through $R5$ are used in this circuit. If the values of $R1$ through $R4$ are greater than or equal to the value of $R5$, the output current is found by the formula

$$\text{Output current} = \frac{\text{Input voltage}}{R5}$$

Changing the input voltage will therefore change the output current. The input voltage can be any value up to ± 15 V; the TL081 uses a dual-polarity supply of up to ± 18 V.

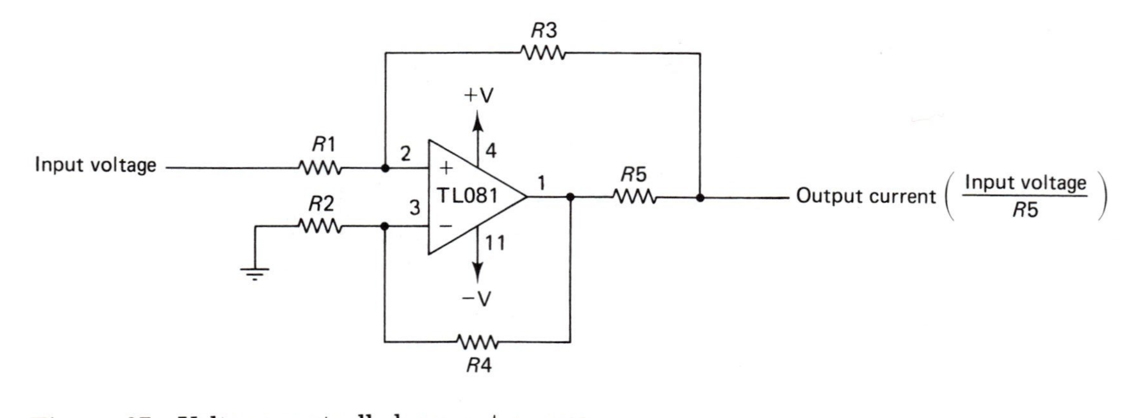

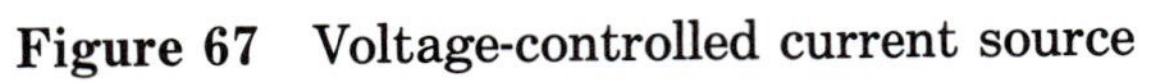

Figure 67 Voltage-controlled current source

d

Data distributors

A data distributor takes a single data input signal and distributes it to one of several different possible outputs depending upon the control signals applied to it. One popular IC for this purpose is the 4051, a switching device which can work with either digital or analog input signals. The circuit shown in Figure 68 is configured for digital input signals. The output selected is determined by the digital inputs to the address inputs, with A = 1, B = 2, and C = 4.

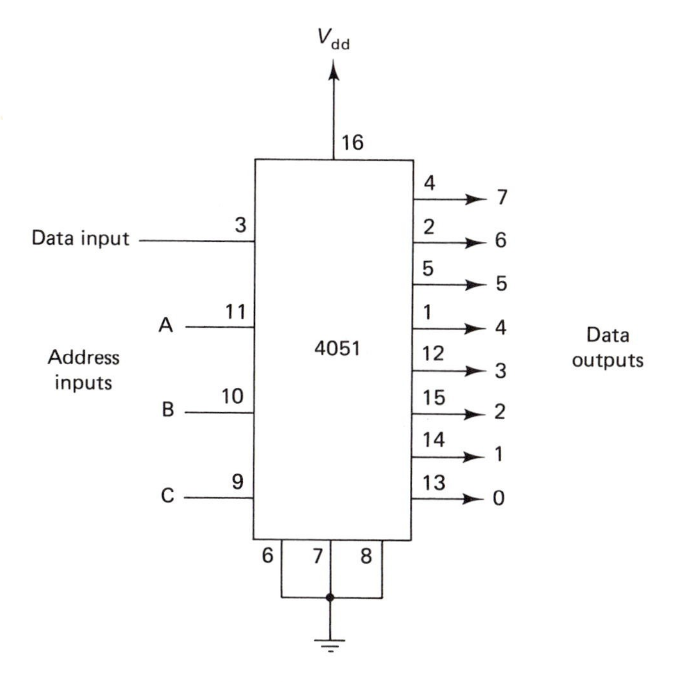

Figure 68 1-of-8 data distributor

For example, if A and C are high while B is low, output 5 is selected; if all are low, output 0 is selected. The digital input and control signals should be equal to the supply voltage and ground. If an analog input is used, pin 7 should be connected to −5 V instead of ground and the analog input kept between −5 and +5 V; the control signals should be +5 V and ground.

Figure 69 shows a 16-output data distributor using the 74154 IC. Unlike the 4051, this device can be used only with TTL-level digital signals. Four address inputs are provided, with A = 1, B = 2, C = 3, and D = 4. With the 74154, the selected output goes low while the remaining 15 stay high.

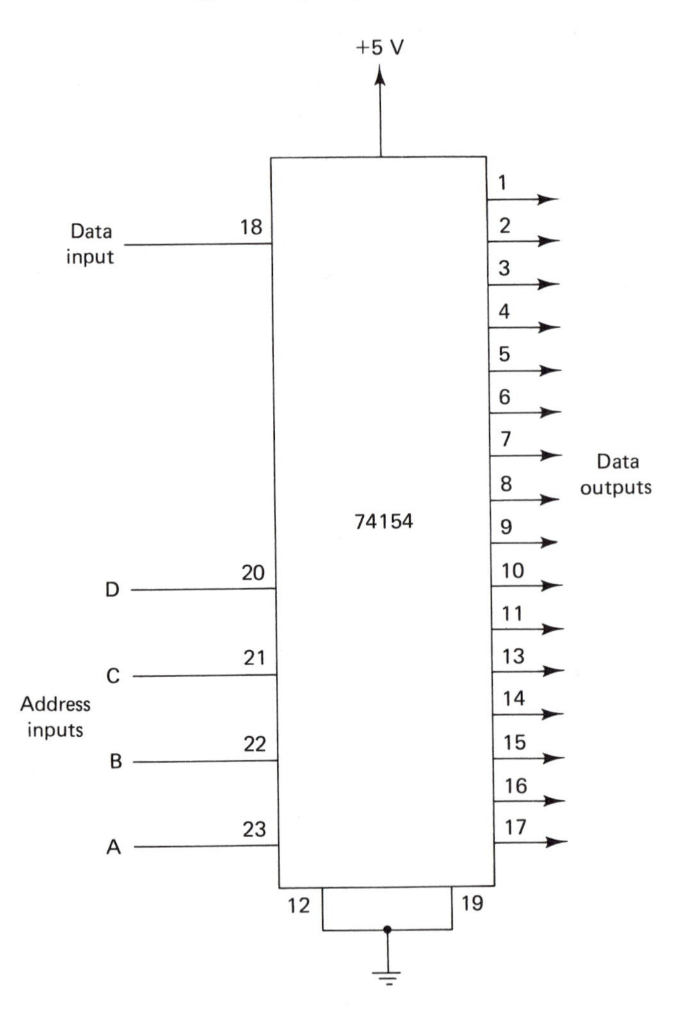

Figure 69 1-of-16 data distributor

Data selectors

The function of a data selector is the opposite of that of a data distributor, selecting one among several input signals and routing it to a single output line.

Figure 70 shows a four-input selector which uses the 4066 quad bilateral switch IC. This circuit can be used with analog or digital signals. Its operation is simple: when a high logic signal is applied to any of the data selection inputs, the corresponding data input is activated and its signal appears at the output. All other data selection inputs must be kept low.

Eight data inputs may be selected with the circuit in Figure 71. This uses the same 4051 device that is used as a data distributor in Figure 68, but this time it is configured as a data selector. The

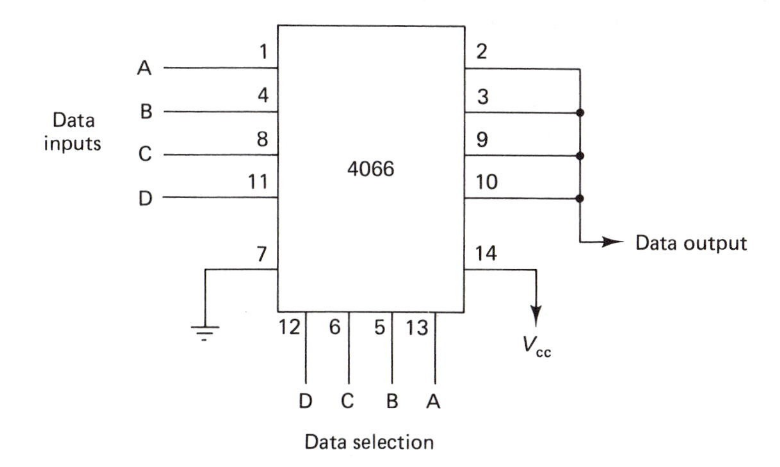

Figure 70 Data selector

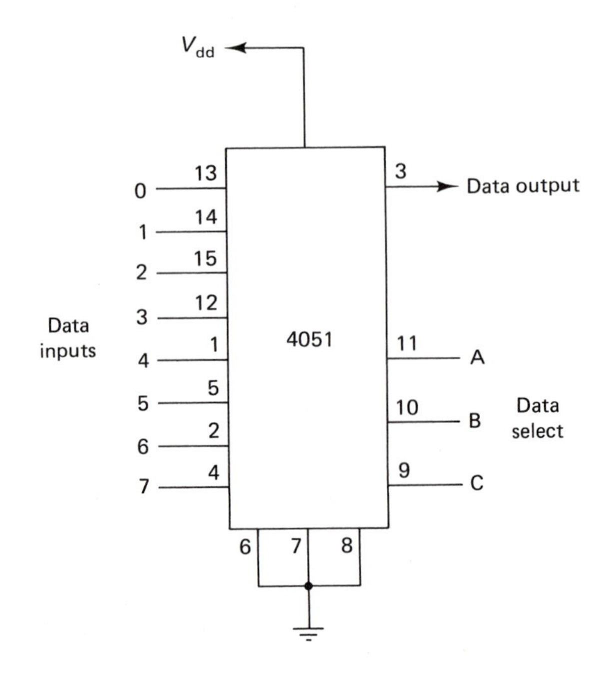

Figure 71 1-of-8 data selector

data select inputs work the same as when the 4051 is configured as a data distributor: A = 1, B = 2, and C = 4. It can also work with analog or digital input signals, as described in the data distributor example in the previous section.

Decoders, FSK

Frequency shift keying (FSK) is a method of transmitting data by switching either an audio tone or radio signal between two fixed frequencies. These frequencies, known as the *mark* and *space,* come in pairs, such as 2,025 and 2,225 Hz. When heard as an audio signal, FSK makes a "tweedling" sound. FSK is widely used in radio and in computer communications; each computer modem is a complete FSK transmitter and decoder.

Figure 72 shows a versatile FSK decoder which can be optimized for different transmission rates and mark/space frequencies. The heart of this circuit is the XR-2211 FSK demodulator–tone decoder IC, which has been designed specifically for FSK demodulation. The internal circuitry of the XR-2211 includes a PLL, phase detector, and voltage comparator. The device, and the circuit in the figure, can operate over a supply voltage range of 4.5 to 20 V.

Figure 72 shows several components with labels instead of parts values. The reason is because these parts are changed for the baud rate and operating frequencies desired. The following is a listing of the appropriate values:

300 Baud, 1,070/1,270 Hz

$C1 = 0.039\ \mu F$ $\quad R1 = 18\ K$
$C2 = 0.01\ \mu F$ $\quad R2 = 100\ K$
$C3 = 0.005\ \mu F$

300 Baud, 2,025/2,225 Hz

$C1 = 0.022\ \mu F$ $\quad R1 = 18\ K$
$C2 = 0.0047\ \mu F$ $\quad R2 = 200\ K$
$C3 = 0.005\ \mu F$

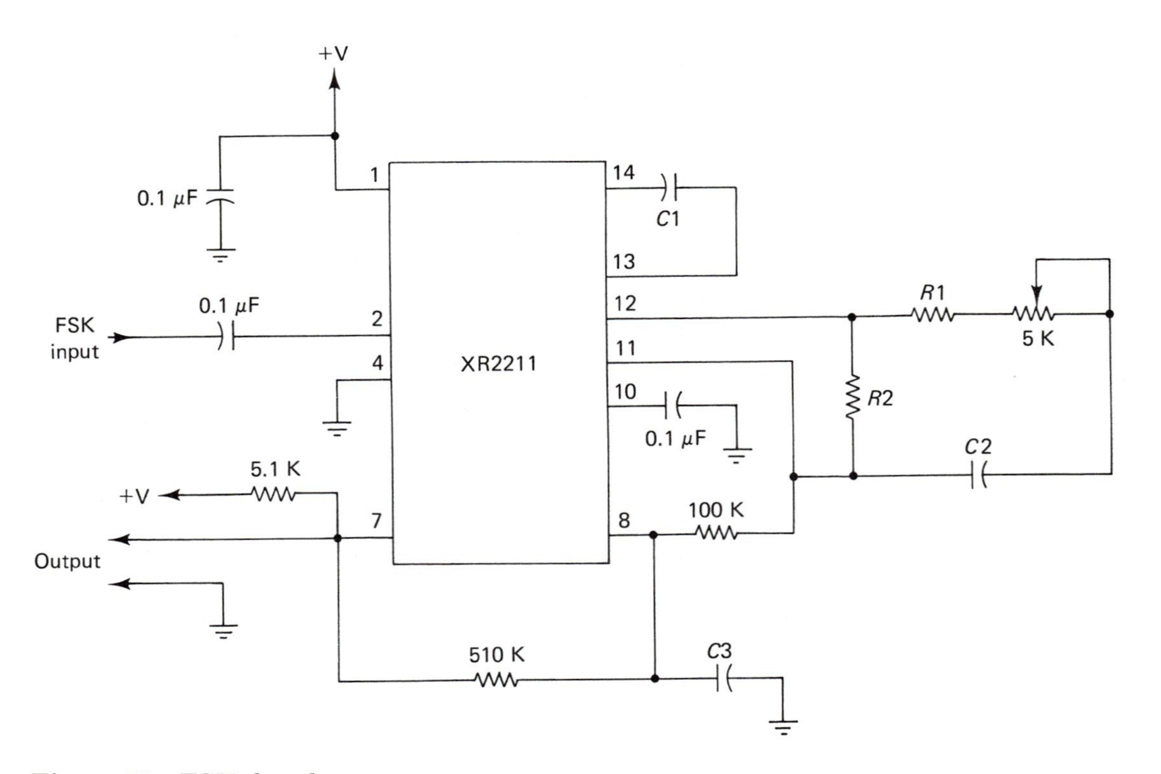

Figure 72 FSK decoder

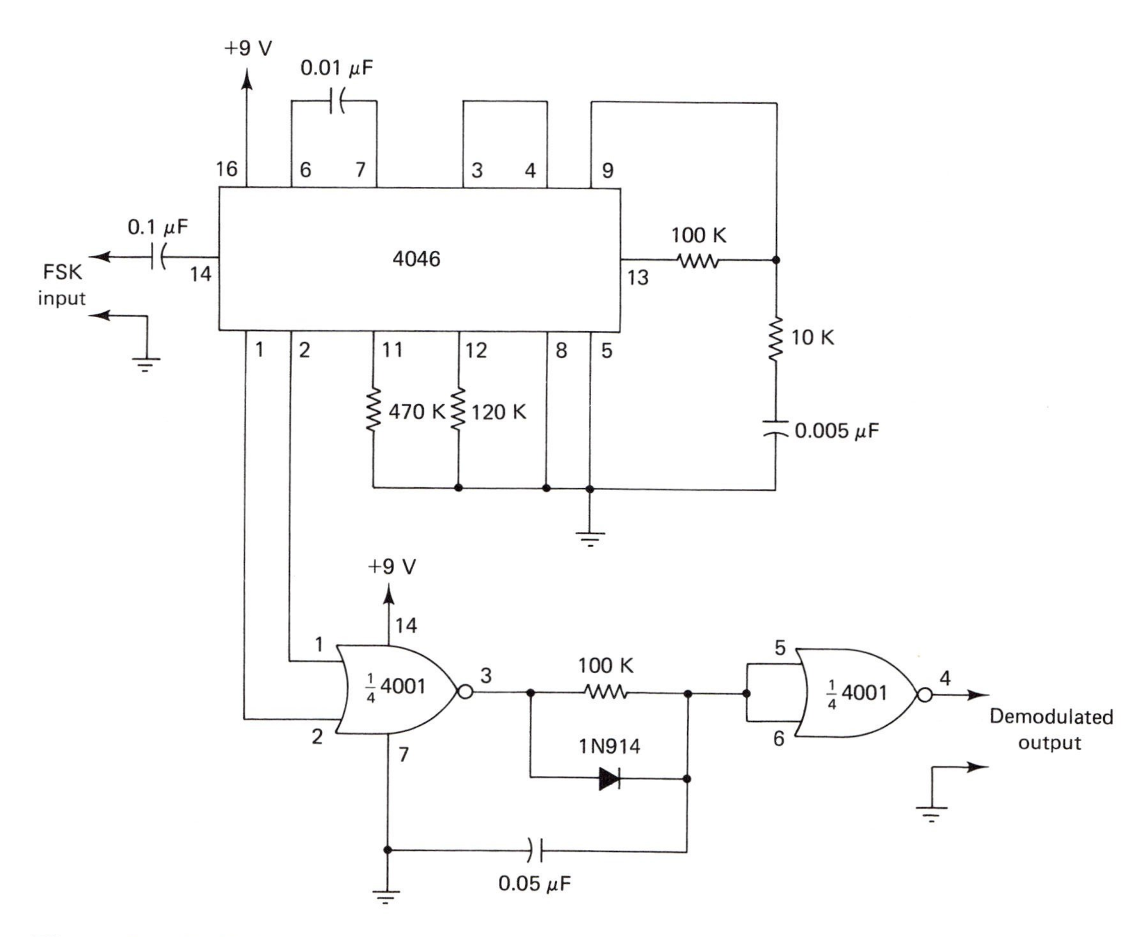

Figure 73 FSK detector

1,200 Baud, 1,200/2,200 Hz

$C1 = 0.027\ \mu F$ $R1 = 18\ K$
$C2 = 0.01\ \mu F$ $R2 = 30\ K$
$C3 = 0.0022\ \mu F$

It is also possible to construct FSK decoders from PLL ICs and supporting devices. Figure 73 shows an "all CMOS" FSK decoder using the 4046 PLL and two gates of a 4001 quad NOR gate IC. This circuit is designed for 300 baud transmission rates at frequencies of 2,100 and 2,700 Hz, the so-called "Kansas City standard" for microcomputers.

Demodulator, FM

The 4046 PLL device can be used in FM demodulators, which take an FM input signal (usually from the IF amplifier stage of an FM receiver) and produce an output suitable for audio amplification in a conventional manner. The circuit in Figure 74 will work with most popular IF frequencies used in FM receivers and deliver usable output, although it can generally be used throughout the frequency range of the 4046.

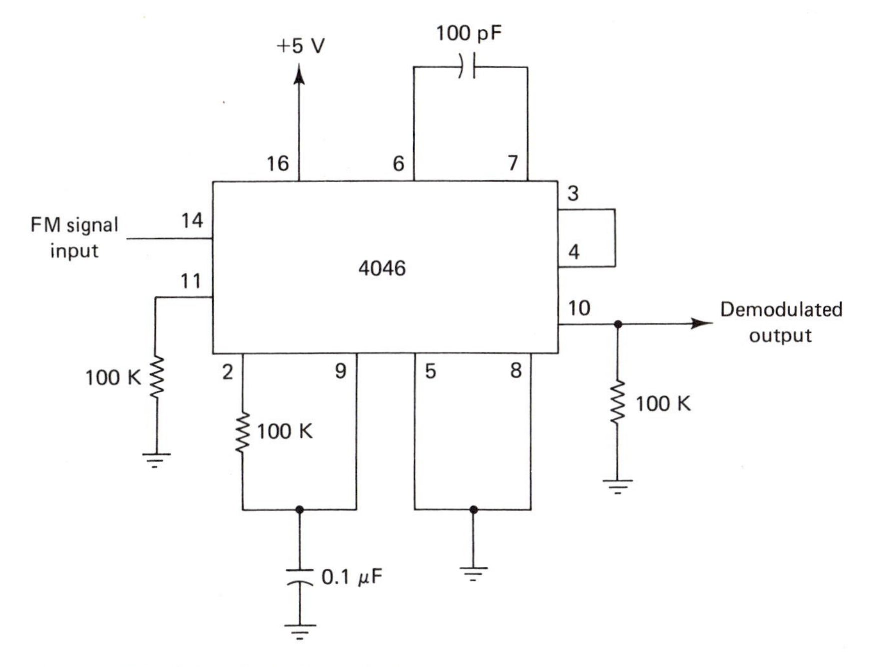

Figure 74 PLL FM demodulator

Detectors, edge

Many digital ICs operate with what is known as *edge clocking,* where the clock signal is defined as the *edge* of a change (from low to high or from high to low) rather than the level of the clock signal. The method of clocking used—whether edge or level—can have important implications in the operation of the device. Thus, it is often important to have some indication of an edge transition.

Figure 75 shows two edge detector circuits for use with TTL circuits. Notice that the input signal for each circuit is defined as

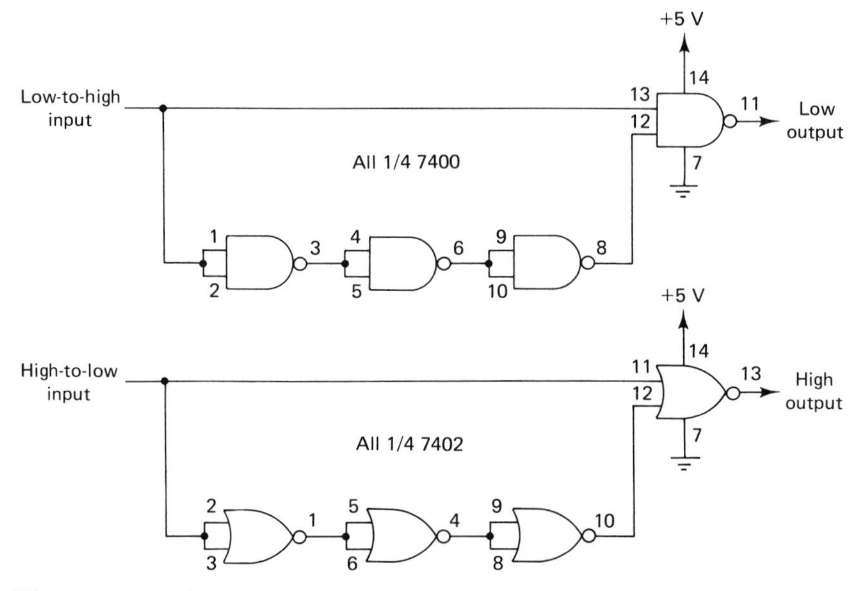

Figure 75 Edge detectors (TTL)

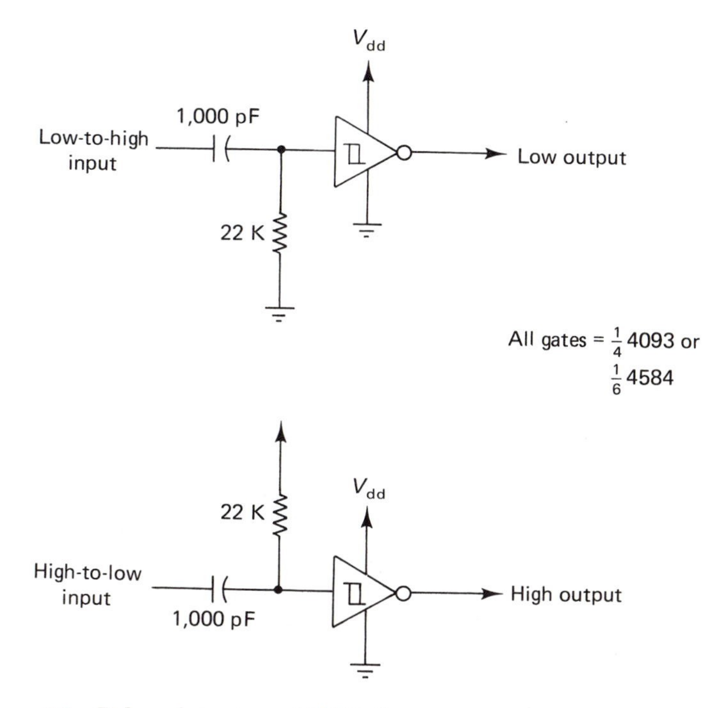

Figure 76 Edge detectors (CMOS)

a transition (as from low to high) rather than a constant level. The output indicated takes place when the transition takes place; otherwise, the output is the opposite of that shown. Figure 76 shows similar circuits built using CMOS and intended for use with CMOS devices. LEDs may be added for visual indication of an edge transition, or the circuits may serve as the inputs to other digital ICs.

Detector, logic level

This circuit, shown in Figure 77, uses one section of a 4050 hex noninverting buffer to give a visual indication of the logic level of a CMOS source. The LED glows when the input signal is low and goes out when it is high. If a reverse function is desired, the output of the 4050 can be followed by an inverter (NOT gate) or by a section of a hex inverting buffer such as the 4049.

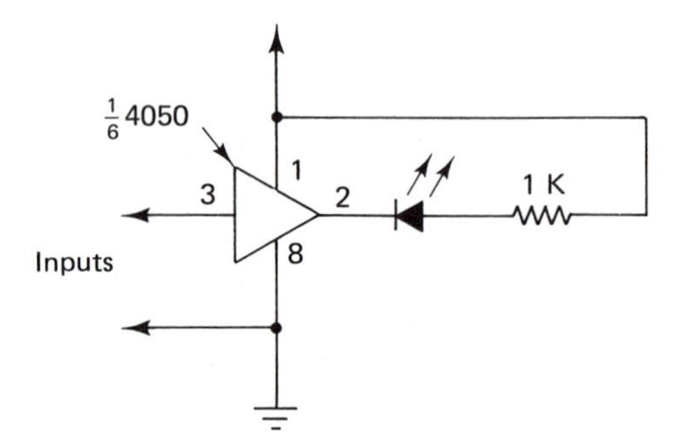

Figure 77 Logic level detector

Detector, missing pulse

This circuit monitors a stream of input pulses, such as those from a clock signal, and maintains a constant high output as long as the stream is uninterrupted. But if the stream "misses a pulse," the output of the circuit goes low for the duration of the missing pulse.

Figure 78 shows a missing pulse detector that is a variation on a monostable multivibrator using a 555 timer device. This IC normally disregards input pulses which arrive during its timing cycle. However, the input pulses bias the MPS2907 transistor into con-

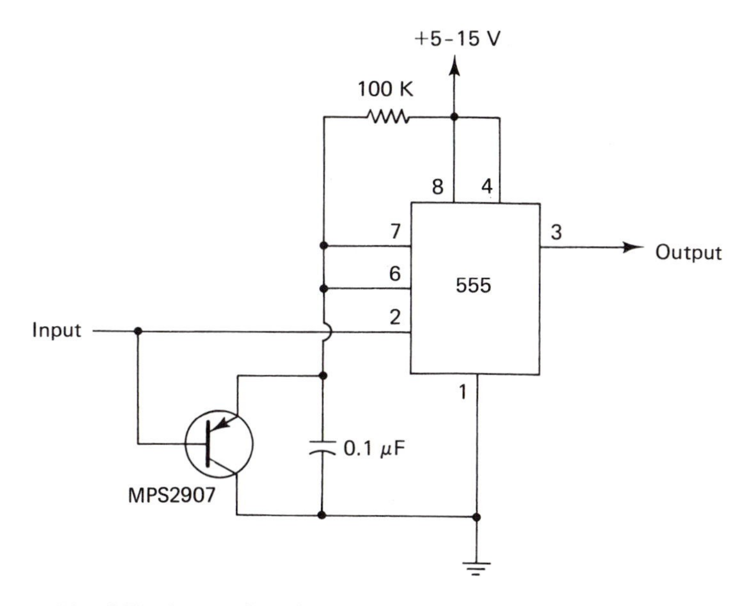

Figure 78 Missing pulse detector

duction, discharging to 0.1 μF capacitor and beginning a new timing cycle. As long as the interval between input pulses is less than the timing period, the output is high. But if a pulse does not arrive until after the previous timing cycle ends, the output goes low until the next pulse arrives. The timing cycle can be altered by varying the values of the 100 K resistor and 0.1 μF capacitor.

Detectors, peak

A voltage detector monitors an input voltage and "stores" the maximum level it reaches using a capacitor. This maximum voltage then becomes the output of the circuit. Figure 79 shows a representative peak detector using both halves of a 1458 dual op amp. The maximum voltage is stored in a 100 μF capacitor. Once it has been charged by an input voltage, the capacitor will not be charged further unless the input voltage increases. The 1N914 diode prevents the capacitor from discharging through the op amp stage and possibly damaging it. The second half of the 1458 is used as a buffer between the peak detector itself and other circuits. The output of this circuit has the same polarity as the input.

Figure 80 shows a peak detector which is resettable by a push-

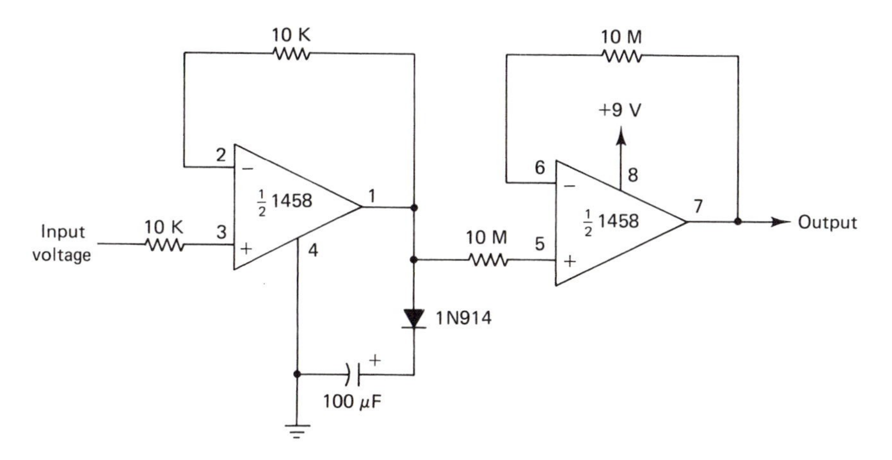

Figure 79 Peak detector

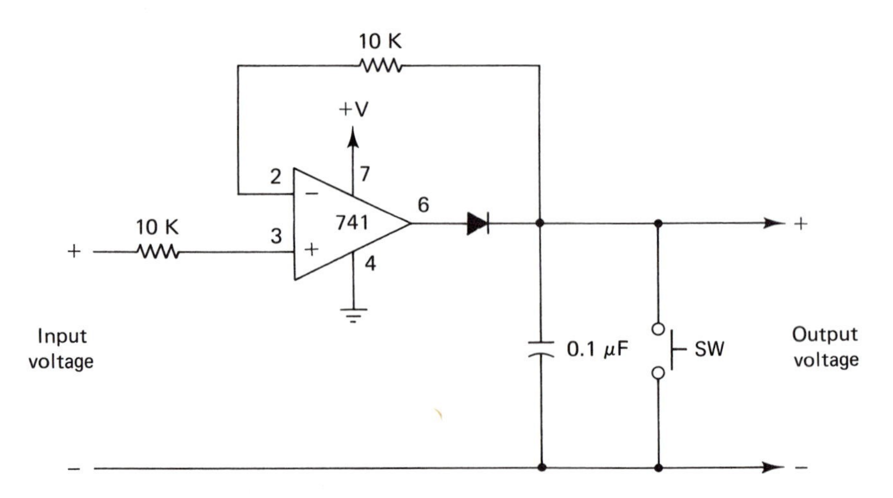

Figure 80 Resettable peak detector

button, SW. This allows the 0.1 μF capacitor to be cleared to zero voltage after the charge it holds has been read. Once the capacitor has been cleared, it can once again begin charging.

Detector, PLL lock

It is often useful—or essential—to know when or if a PLL is "locked." Consider, for example, the circuit of Figure 81, which was designed to work with a 4046 CMOS PLL IC but should work with most other CMOS PLL devices. The circuit consists of the four NOR gates on a 4001 quad NOR gate IC. The inputs are from pins 1 and 2 of the 4046, and prior to that, from the phase pulse outputs of the two phase comparators on the 4046. When the PLL is locked on an input signal, the output of this circuit will be high; when it is not locked, the output is low. A LED or similar visual indication of loop lock may be added to the output of the circuit.

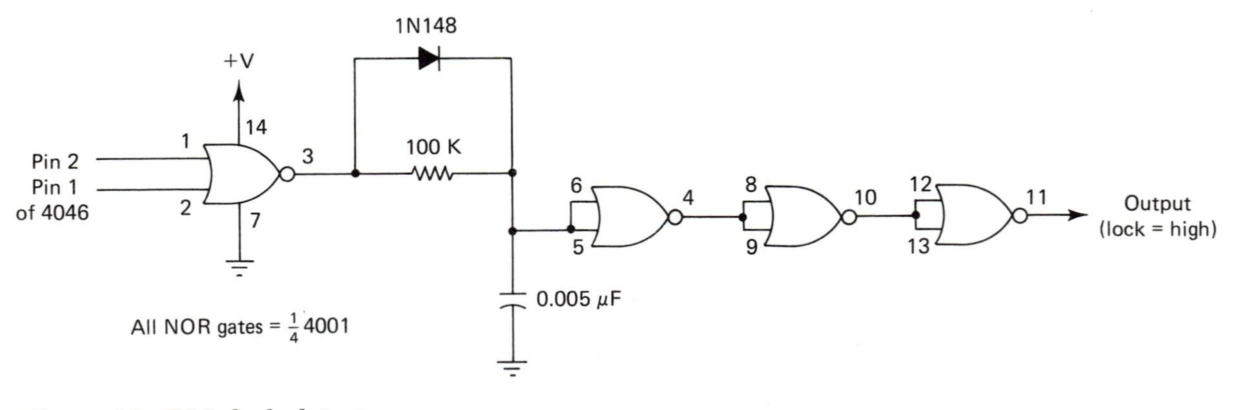

Figure 81 PLL lock detector

Detector, product

A product detector, used for reception of single-sideband signals, operates by combining the received SSB signal with a replacement carrier to allow demodulation. Figure 82 shows a product detector that uses a single IC. The heart of this circuit is an MC1496 balanced modulator and demodulator, in which the output voltage is a product of the input voltage signal and a switching function supplied by the carrier. The circuit has a sensitivity of 3 μV and a dynamic range of 90 dB when operating at an IF frequency of 9 MHz. It is broadband for the entire high-frequency range. The replacement carrier may be derived from the IF signal or generated separately.

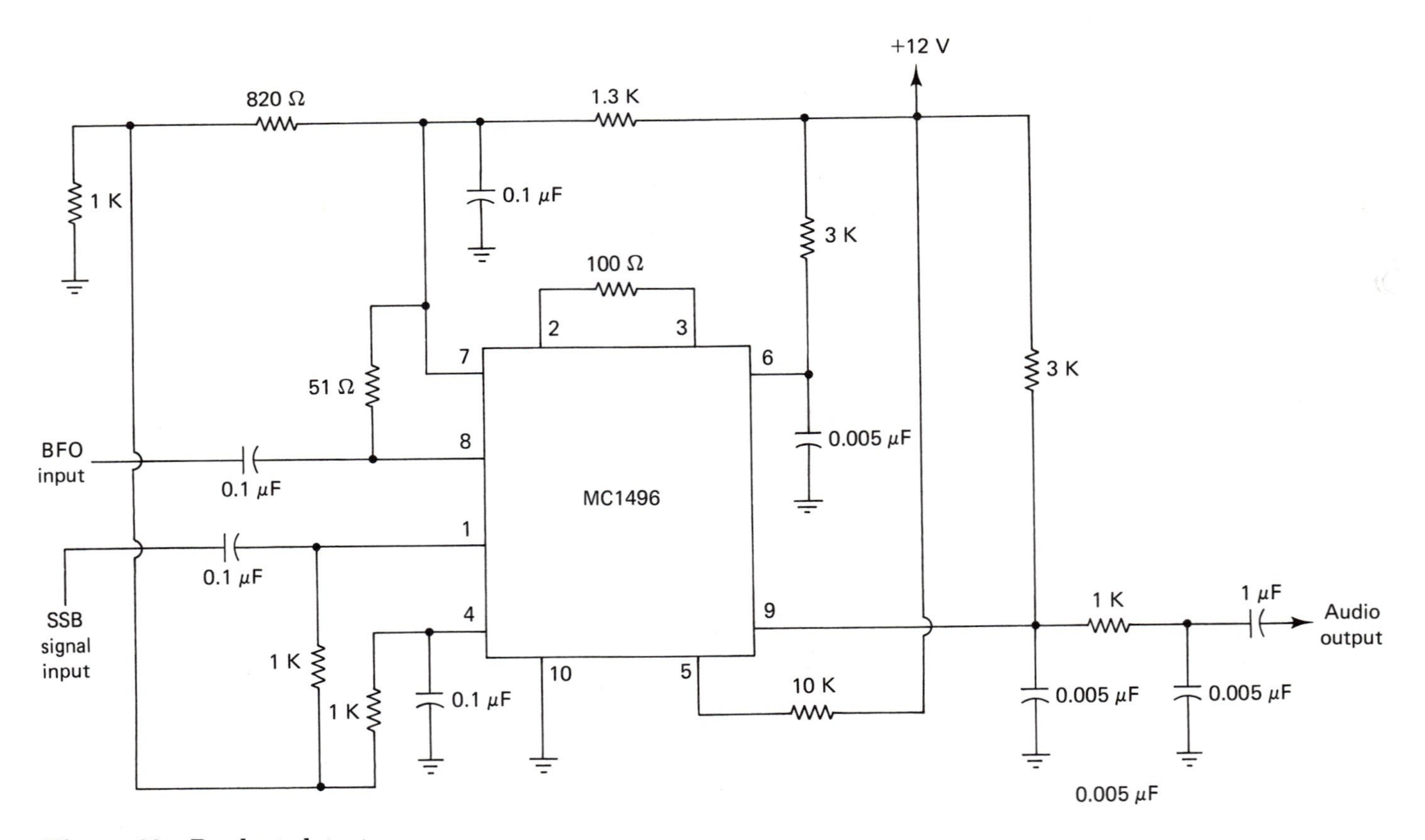

Figure 82 Product detector

Detector, voltage level

A voltage level detector is a bit like a comparator: it produces an output only when an input voltage exceeds a set level. The circuit in Figure 83 will light a LED when the input voltage exceeds a level set by the 50 K potentiometer. The input voltage must not exceed the supply voltage; indeed, it is good practice to keep the input voltage slightly below the supply voltage.

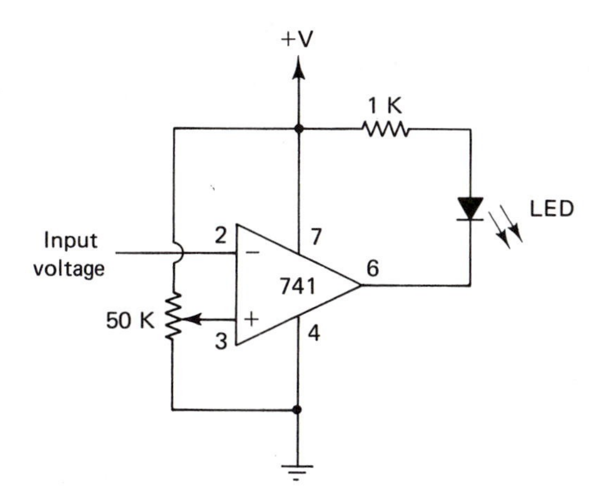

Figure 83 Voltage level detector

Detectors, zero crossing

A zero crossing detector is a circuit whose output changes each time the input signal crosses its base line (the zero point of the input waveform). A common use for these circuits is to produce square wave digital signals from an audio input, as, for example, in loading a microcomputer program from an audio cassette. Figure 84 shows a 741 op amp used as a zero crossing detector with a TTL-level output; the latter is why the supply voltage must be held to 5 V. The circuit in Figure 85 shows a similar circuit with a CMOS-compatible output. The 3 K potentiometer between pins 5 and 6 should be adjusted to provided a zero output signal when no input signal is applied.

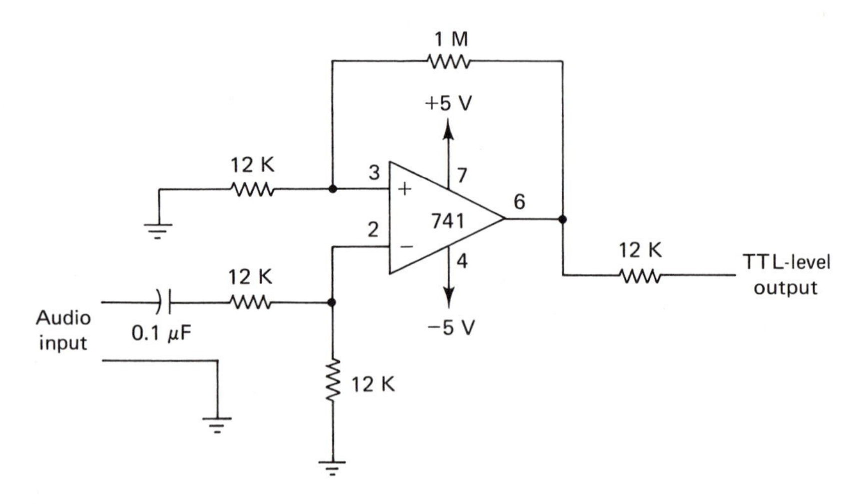

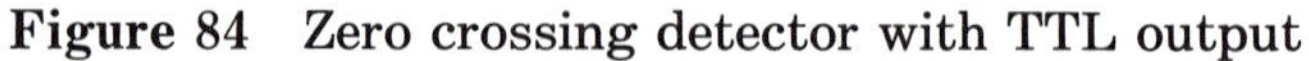

Figure 84 Zero crossing detector with TTL output

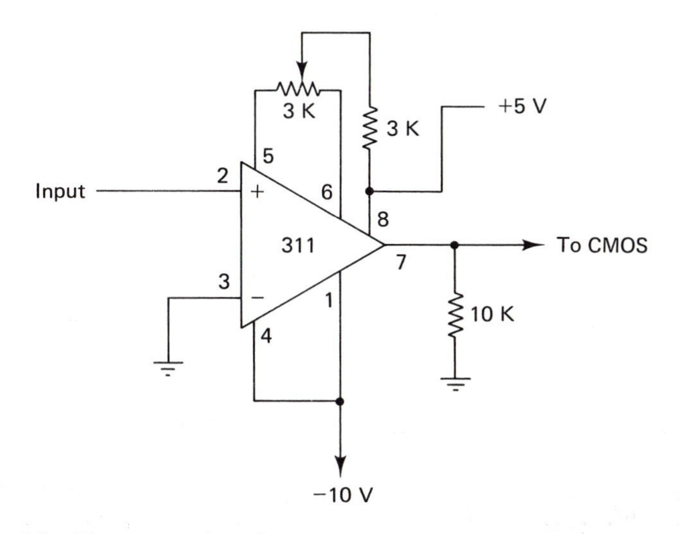

Figure 85 Zero crossing detector with CMOS output

Detector, 2,600 Hz tone

The frequency of 2,600 Hz is widely used in telephone systems for controlling signaling between exchanges. When the tone is present it indicates that a telephone is "on hook," and when it is not the telephone is "off hook." Thus, the accurate detection of the presence or absence of this signal on a telephone line is crucial for proper operation of the system. The circuit in Figure 86 shows an S3526B programmable low-pass filter combined with an S3524A digital frequency detector. The result is a circuit with a center frequency of 2,600 Hz and a bandwidth of 70 Hz. The internal reference frequency is controlled by a standard 3.58 MHz "color burst" crystal. The output of this circuit will high as long as a 2,600 Hz tone within the 70 Hz bandwidth is detected from the input. When the tone ends or drifts out of the specified bandwidth, the output goes low.

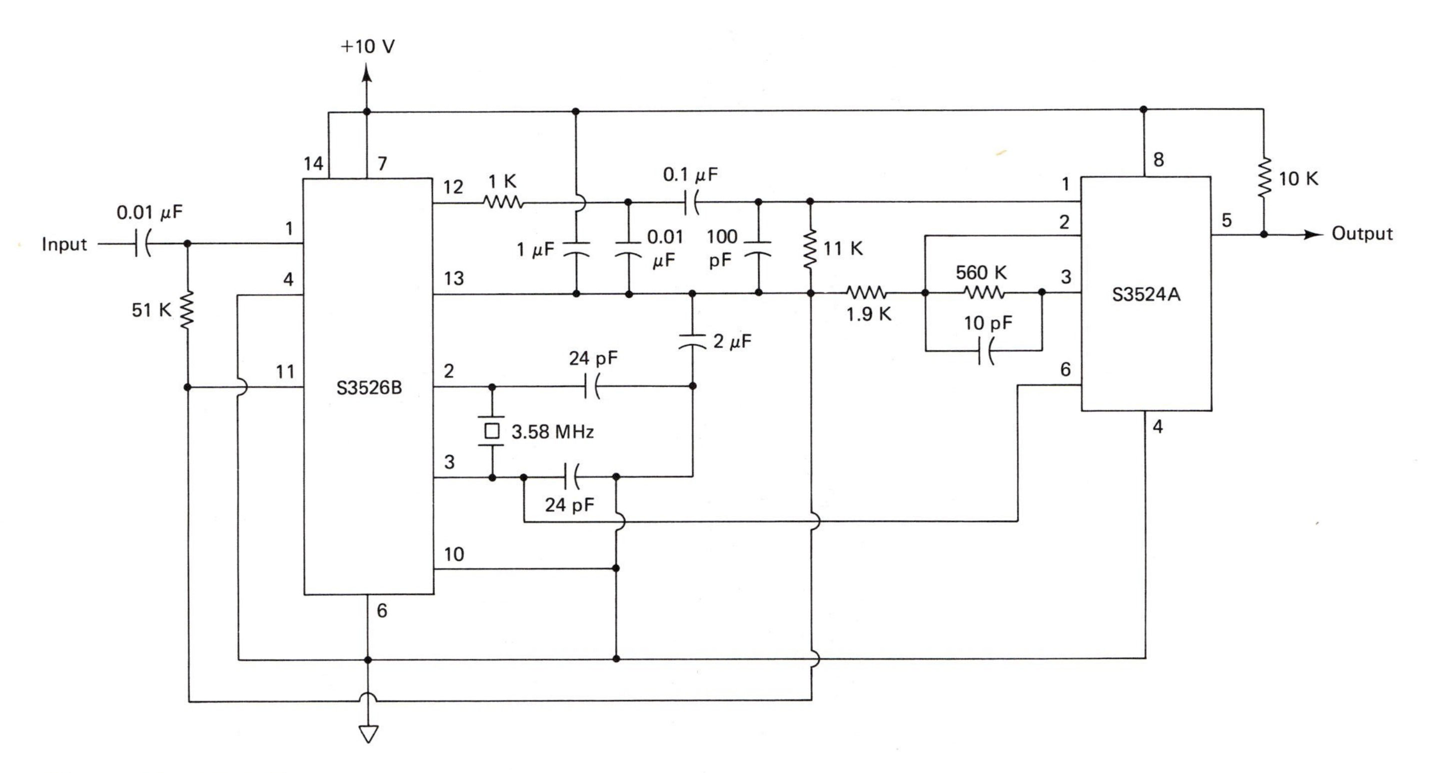

Figure 86 2,600 Hz tone detector

Differentiator

A differentiator accepts a triangular waveform input and produces a square wave output. (It can also accept a square wave input and produce a series of "spike" pulses.) In more technical terms, this can be described as producing an output voltage that is proportional to the rate of change of the input voltage. The input frequency a differentiator can accept depends upon the values of components $R1$, $R2$, and C shown in Figure 87. The value of $R1$ is approximately one-tenth the value of R2, or

$$R1 = \frac{R2}{10}$$

The value of C is determined by the reciprocal of the value of $R2$ multiplied by the input frequency, or

$$C = \frac{1}{R2 \times \text{input frequency}}$$

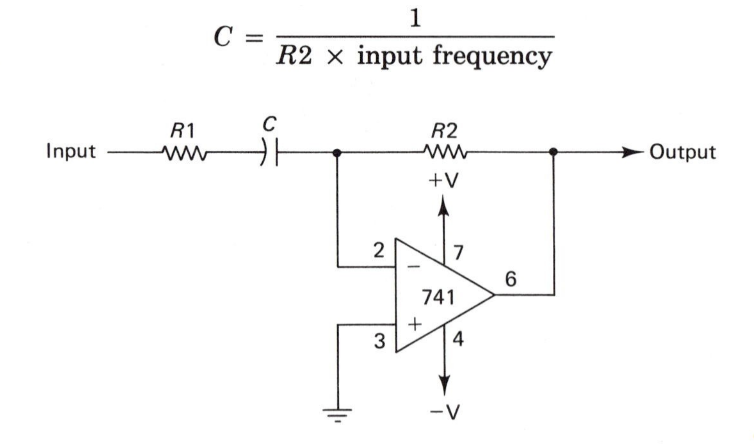

Figure 87 Differentiator

Similarly, the value of $R2$ can be found if the value of C is known by the formula

$$R2 = \frac{1}{C \times \text{input frequency}}$$

Divider, voltage

The circuit shown in Figure 88 takes two input voltages, X and Y, and produces an output voltage given by

$$\text{Output Voltage} = + \frac{10\,X}{Y}$$

The circuit uses an XR-2228 monolithic multiplier IC. Input voltage X can be of either polarity, but input voltage Y *must* be negative; positive values of Y will cause "latchup" of the circuit. (This latchup will not damage the XR-2228, however). The various potentiometers are used to adjust the circuit for most accurate operation, and calibration should be done with known reference voltages.

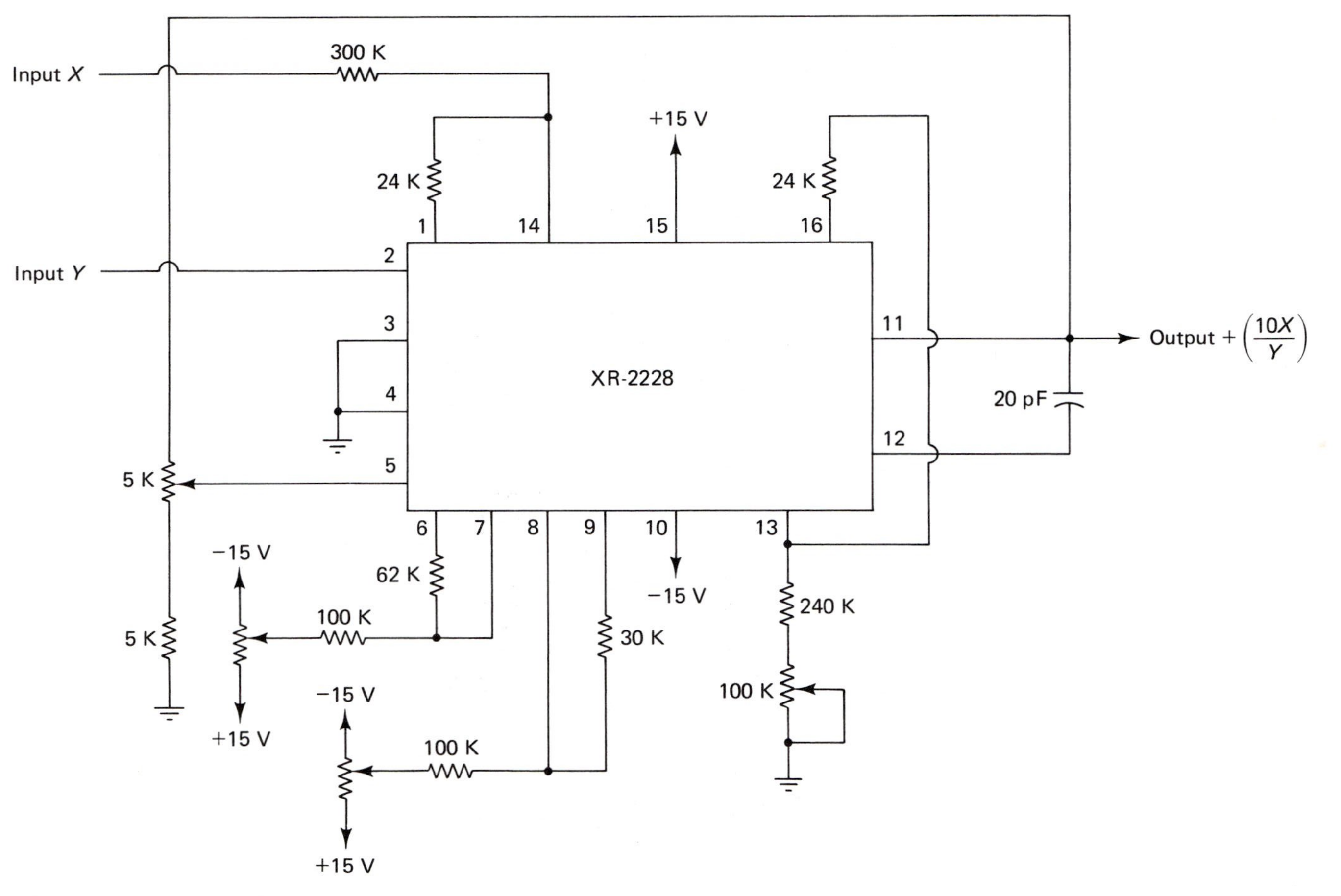

Figure 88 Voltage divider

Doubler, 150-to-300 MHz

A frequency doubler is a circuit in which the frequency of an input signal is doubled while the signal waveform itself is unaffected. Figure 89 shows a frequency doubler that uses an MC1596 balanced modulator/demodulator IC which takes a 150 MHz input and produces a 300 MHz output. The MC1596 has two frequency inputs (at pins 1 and 4), and the same 150 MHz input is applied to both. The 50 K potentiometer is adjusted so that both inputs receive the same signal level, and the two 1–10 pF variable capacitors are adjusted until the resulting output is exactly twice the input and waveform distortion is at a minimum. The input and output impedances are both approximately 50 Ω, the input signal level is typically 100 mV RMS, and the output is about 15 mV RMS. As is normal in circuits operating at these frequencies, careful placement of parts and an adequate ground are essential for proper operation.

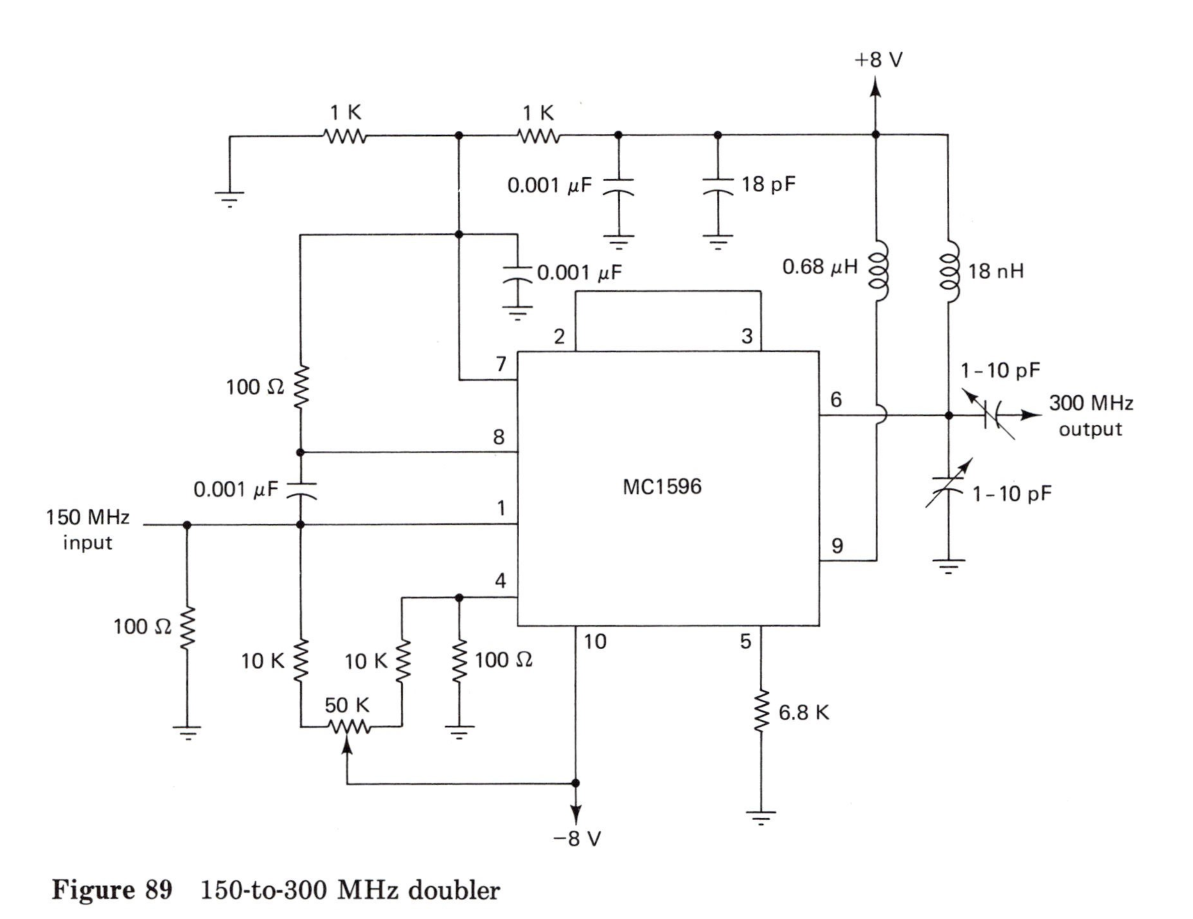

Figure 89 150-to-300 MHz doubler

e

Expander

The audio expander circuit shown in Figure 90 reverses the effects of the audio compressor mentioned earlier. An expander using a PLL such as the NE571 requires a sufficient input signal for a reference to establish a threshold signal level which the input must exceed for proper operation. As long as the received signal is a few dB over noise, the expander will restore the signal to its original form. Otherwise, it will not function correctly. This circuit, when used in conjunction with the compressor circuit, can produce a 12 to 15 dB improvement in signal-to-noise ratio.

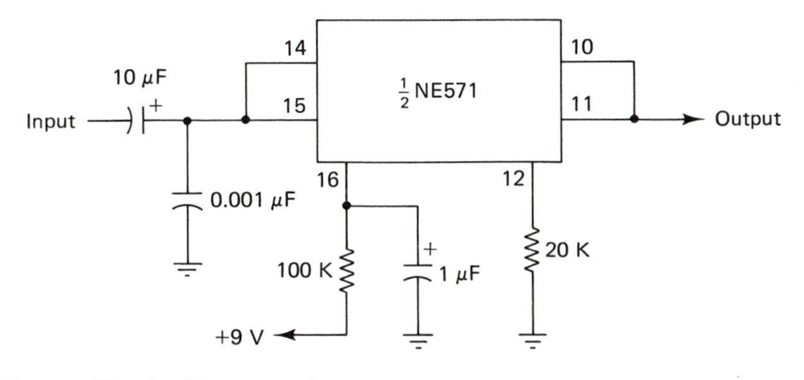

Figure 90 Audio expander

f

Field strength meter

A transmitter or oscillator circuit whose output frequency is up to 150 MHz can be adjusted for maximum output using the circuit shown in Figure 91. This field strength meter is basically a diode detector whose output is amplified until it can drive a meter. The circuit is broad in frequency coverage, but relatively insensitive; for best results, the meter should be placed immediately adjacent to the circuit being tested. To use the meter, switch on the oscillator or transmitter circuit and adjust the 1 M potentiometer until a maximum reading is obtained from the meter. Then adjust the oscillator

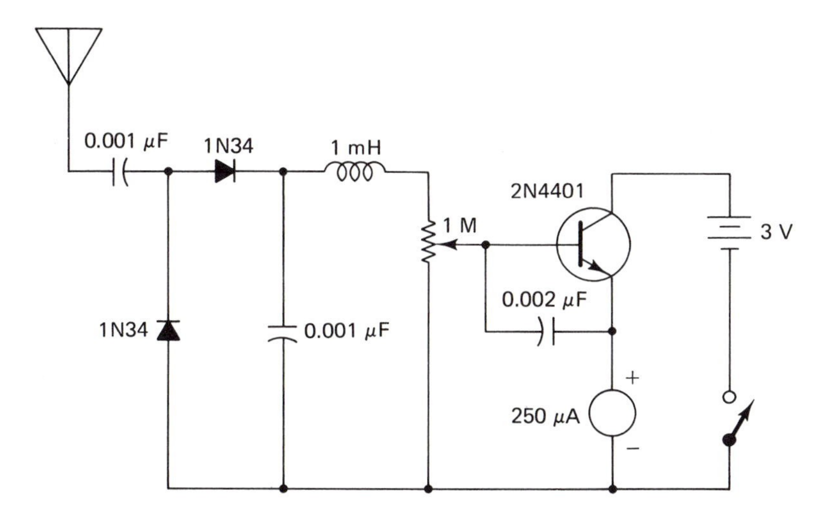

Figure 91 Field strength meter

or transmitter until a new, higher maximum meter indication is produced. The antenna used with this circuit can be any standard telescopic "whip" (such as that used with walkie-talkies or portable radios), and the circuit should be enclosed in a shielded, metal box.

Frequency doubler

Doubling an input frequency can be done with a variety of techniques, such as using an amplifier optimized for harmonic outputs. A simple way to do this at low-level sine waves is with an XR-S200 multifunction PLL device which has an on-chip multiplier section. The circuit in Figure 92 accepts a 10 kHz input at 4 V peak to peak and produces an output of 20 kHz at 1 V peak to peak. The two 10 K potentiometers are adjusted to produce minimum harmonic content.

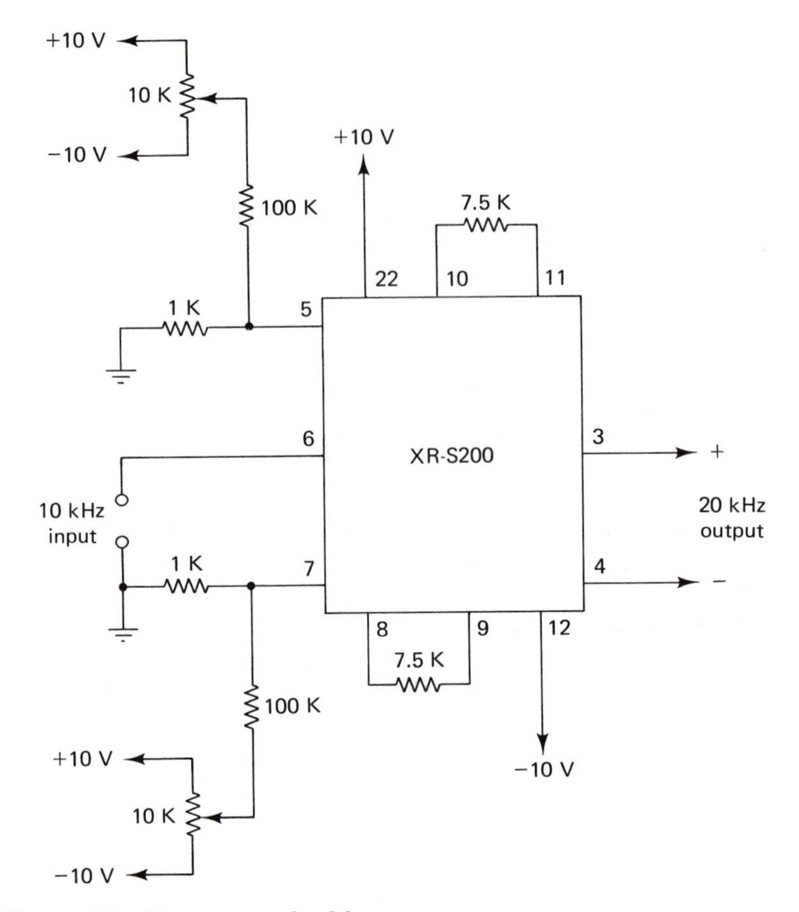

Figure 92 Frequency doubler

Frequency switch

This circuit accepts an AC input signal of 1 V peak to peak and, depending upon the values of $C1$ and $C2$, will light or extinguish the LED depending on whether the input signal is above or below a "trip" frequency. When the LED is on, the input signal is less than the trip frequency; when the LED is off, the input signal is above the trip frequency. The trip frequency is found by the formula

$$\text{Trip frequency} = \frac{1}{25{,}000\ C1}$$

The value of $C2$ depends on the value of $C1$ and is given by

$$C2 = 100C1$$

The circuit in Figure 93 makes use of an MC3524 power supply supervisory circuit and dual comparators. The MC3524 device has an internal 2.5 V reference for the comparators.

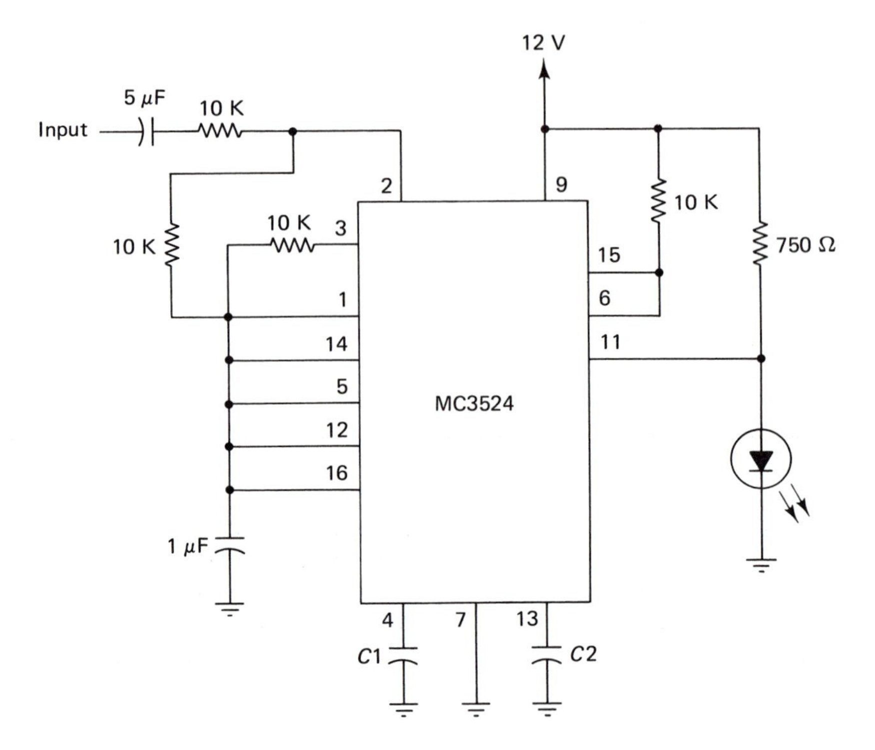

Figure 93 Frequency switch

g

Gated tone source

One interesting aspect of CMOS logic gates is that, unlike TTL and other logic families, they can be used in some "linear" applications such as audio oscillators. Figure 94 shows a 1,000 Hz audio oscillator which uses two sections of a 4001 quad NOR gate IC. Since the output tone is produced by the rapid switching of the NOR gates between states, this circuit is known as a gated tone source. The advantage of this approach is that the circuit can be more stable than many linear audio oscillators of comparable simplicity.

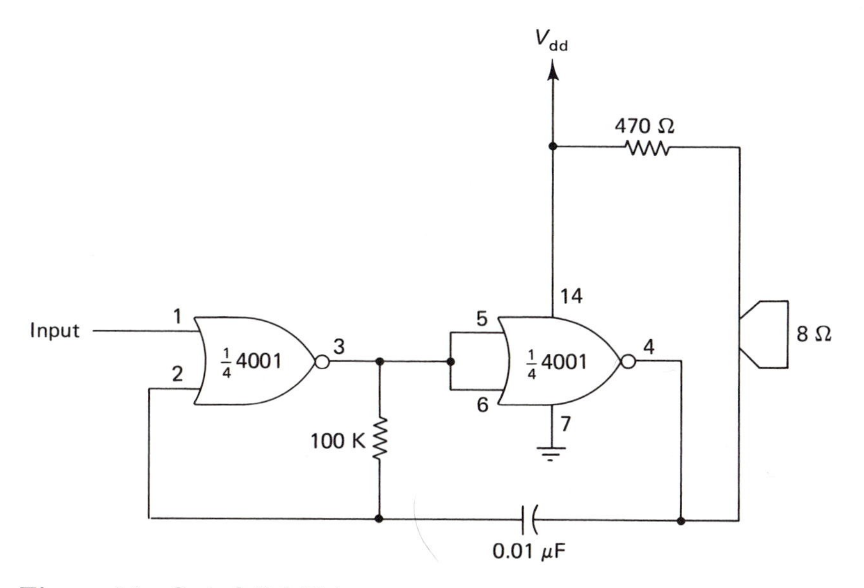

Figure 94 Gated (1 kHz) tone source

i

Integrator

The function of an integrator circuit is essentially the opposite of that of a differentiator. The formal definition of an integrator is that it is a circuit which produces an output voltage proportional to the integral of an input voltage. In practice, this means an integrator takes a square wave input and produces triangular waves at the output. Figure 95 shows a typical integrator. The input frequency which an integrator can accept depends upon the values of $R1$, $R2$, $R3$, and C. The formulas for determining these values are as follows:

$$R1 = \frac{R2}{10}$$

$$C = \frac{1}{R2 \times \text{input frequency}}$$

$$R2 = \frac{1}{C \times \text{input frequency}}$$

$$R3 = \frac{R1R2}{R1 + R2}$$

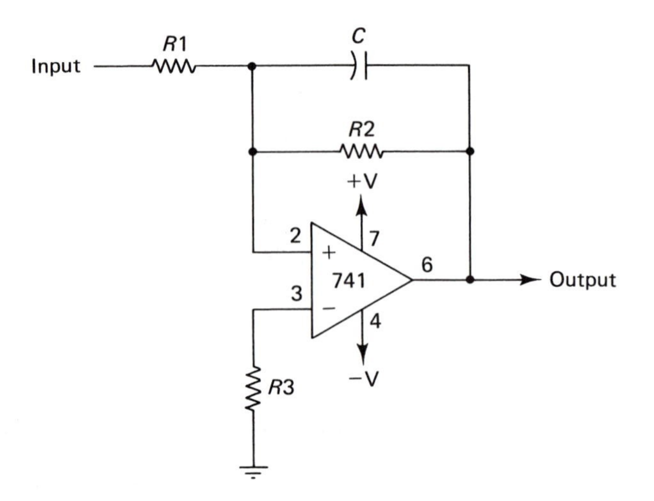

Figure 95 Integrator

Interfacing, digital-to-digital

The two most popular digital logic families in use today—TTL and CMOS—are not directly compatible with each other (i.e., the output of one family cannot serve as the input for the other). Because of this, good design practice tries to avoid mixing the two families in the same circuit. However, sometimes this is unavoidable and some method of enabling the different logic families to "talk" to each other must be provided.

The two major incompatibilities between CMOS and TTL involve supply voltages and output currents. TTL requires a constant +5 V supply voltage, while CMOS can be powered from +3 to 15 V. At first, it might seem a simple matter to power both CMOS and TTL from +5 V. This can be, and often is, done; however, many of CMOS's advantages (e.g., immunity to noise) are lost at lower supply voltages. In terms of output currents, TTL is capable of

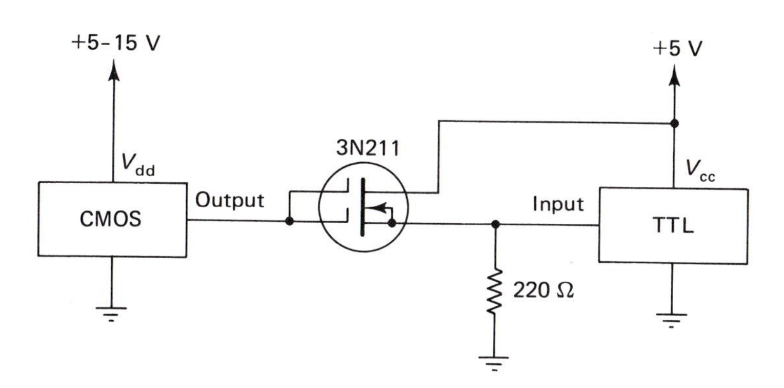

Figure 96 Interfacing CMOS to TTL

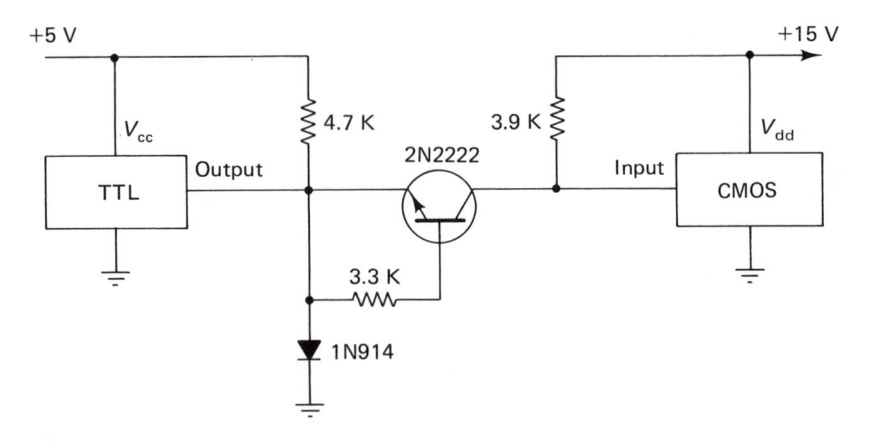

Figure 97 Interfacing TTL to CMOS

"sinking" much more current than CMOS. Thus, the preferred solution in most applications is to provide separate supply voltages for TTL and CMOS and to use a transistor to overcome the differences in output currents. Figure 96 shows how this is done with CMOS-to-TTL interfacing, and Figure 97 shows a similar approach with TTL-to-CMOS interfacing.

Interfacing, linear-to-digital

Most linear devices can accept inputs from digital ICs as long as those inputs are within the voltage and current restrictions for the linear device's inputs. Moreover, several linear ICs—such as the 555 and 556 timers, ICL8038 waveform generator, and 565 PLL—have TTL-compatible outputs. For other linear devices, however, some form of interfacing is necessary before they can drive digital ICs.

Often a comparator will be the linear device supplying the input. A comparator operates much like a digital device, since its output is off until there is a difference between its input signals, at which point it switches to full output. A commonly used comparator IC is the 339 quad comparator. Figure 98 shows how to use this device to provide a CMOS output, while Figure 99 shows how to obtain a TTL output from it. In the first case, the supply voltage must be the same as that used by the CMOS device being driven.

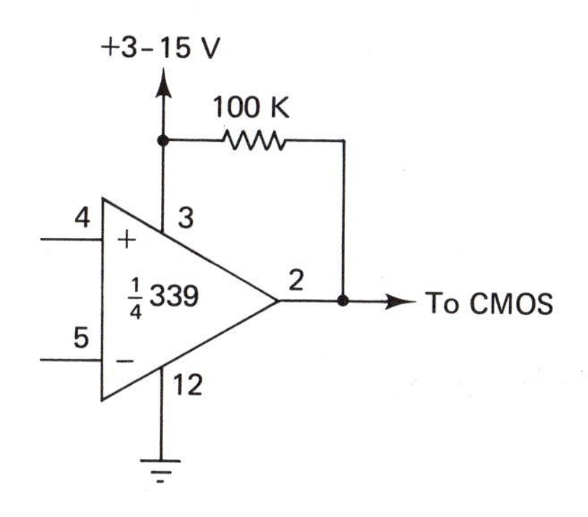

Figure 98 Driving CMOS from an Op amp (comparator)

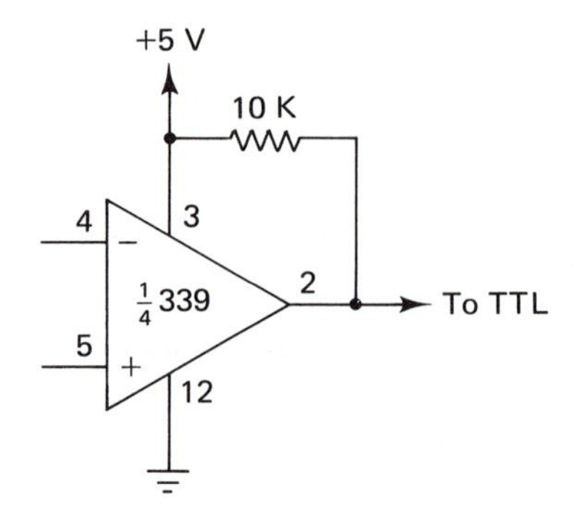

Figure 99 Driving TTL from an Op amp (comparator)

In the second, the comparator and TTL device both operate from a standard +5 V TTL supply.

Logic ICs can also be driven easily from op amps and other linear devices. Figure 100 shows how this can be done. For a TTL input signal, a resistor in the range of 3.3 to 4.7 K connected to +5 V, known as a pull-up resistor, is used to boost the linear output to a level sufficient to drive TTL. For CMOS, all that is necessary is a 10 K resistor in series between the linear output and the CMOS input. This resistor limits current to the CMOS device to a safe level.

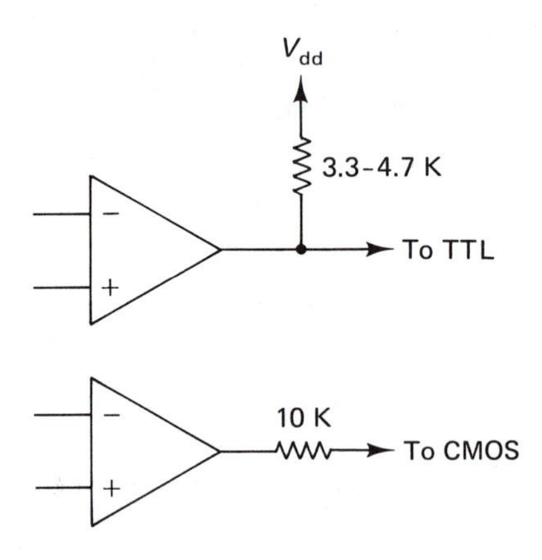

Figure 100 Driving logic from Op amps

Inverter, voltage

A voltage inverter is a circuit which can reverse the polarity of a voltage applied to it; the most common need for such a circuit is to convert a voltage of positive polarity into one of negative polarity. While no ICs specifically intended for this purpose have been designed, it is possible to adapt ICs to perform voltage inversion. Figure 101 shows a circuit in which a μA78S40 switching regulator

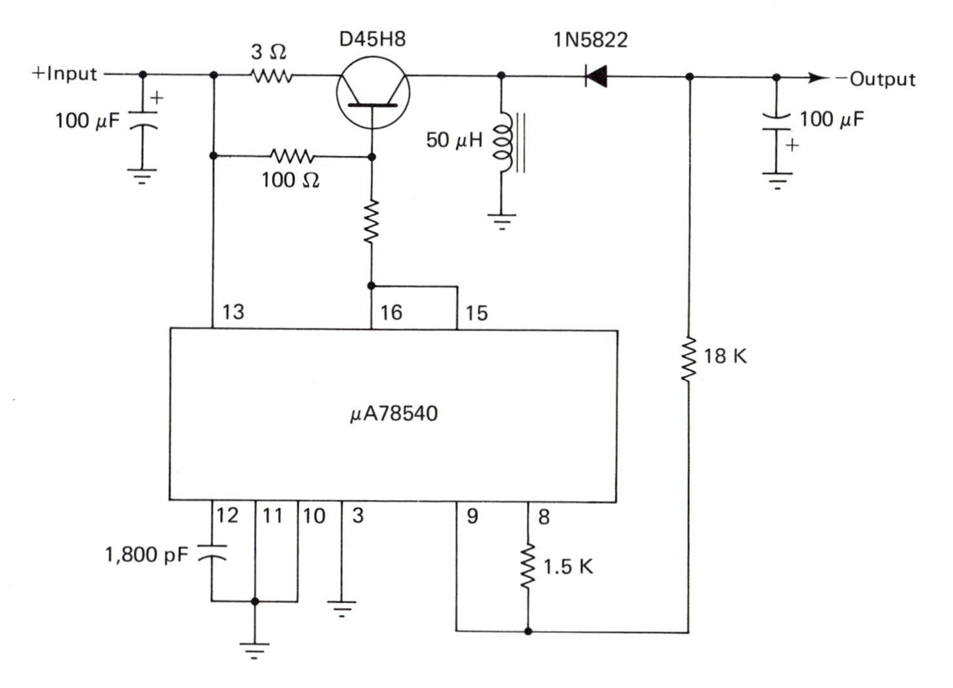

Figure 101 Voltage inverter

subsystem device has been configured to invert an input voltage and deliver substantial output current as well. For example, a +15 V input will cause the circuit to deliver a −15 V output at 500 mA. The circuit is capable of operating with inputs from 2.5 to 40 V. The supply voltage for the IC is obtained from the input voltage at pin 13, and the amount of output current will vary with the input voltage.

1

Latch

Latch circuits are used to hold data output signals even if data input signals change. The operation of the latch is controlled by external signals which tell it whether to accept or ignore the data input signals.

A simple latch is shown in Figure 102. The circuit uses a 74196 counter/divider IC to implement a four-bit latch. When the load signal is low, changes in the data inputs produce corresponding changes in the data outputs. When the load signal is high, the data outputs do not change regardless of any changes in the data inputs.

ICs specifically designed as latches have been developed. Figure 103 shows one such IC, the 7475. When the latch enable signal is high, changes in the data inputs produce corresponding changes in the data outputs. When the latch enable signal goes low, the signals at the data outputs are "frozen" until the latch enable returns

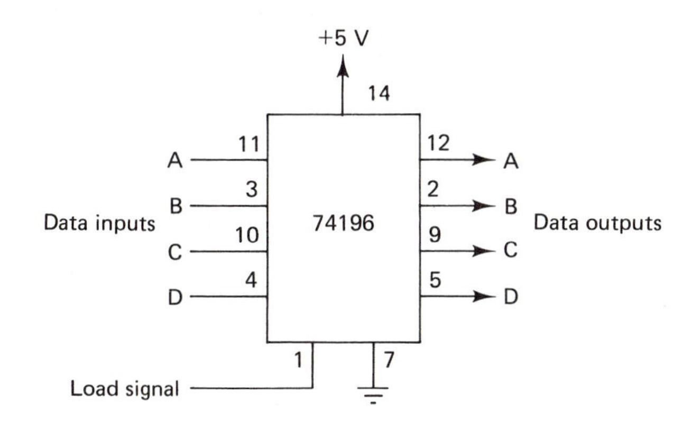

Figure 102 Four-bit latch

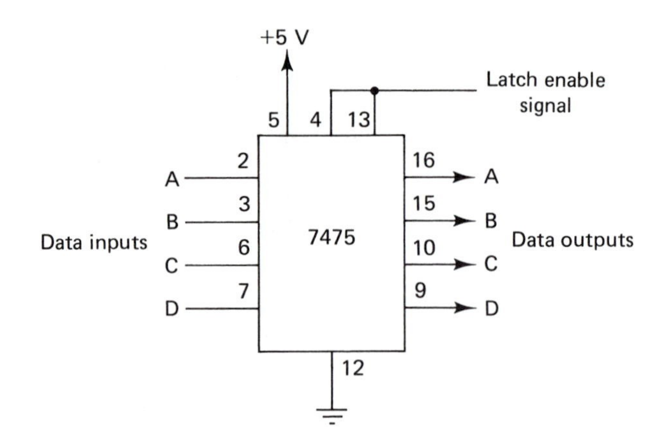

Figure 103 Four-bit latch

to high. Two of these circuits can be combined to create an eight-bit latching sharing a common latch enable signal.

A CMOS device designed for use as a latch is the 4042. Figure 104 shows a circuit in which latch operation is controlled by a clock signal and a clock polarity switch. The clock polarity switch determines which latching action takes place when the clock signal goes

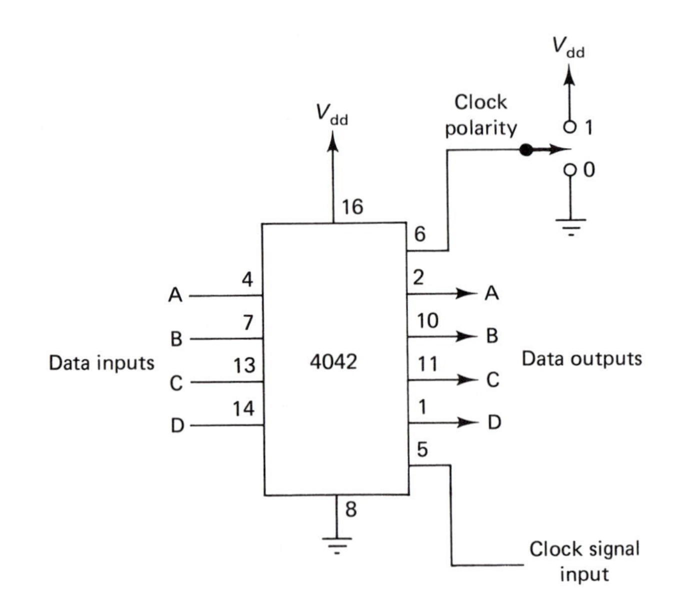

Figure 104 Clocked latch

from low to high or when it goes from high to low. If the clock signal does not change, then regardless of whether it remains high or low, changes in the data inputs produce changes in the data outputs. If the clock polarity switch is set to 0, the circuit latches when the clock signal goes from low to high. If the polarity switch is set to 1, it latches when the clock goes from high to low.

LED flasher

LEDs may be flashed by a variety of circuit configurations. The circuit in Figure 105 uses a 3909 IC, which has been specifically designed for "LED flasher" applications. The flash rate is 1 Hz (60 times per second), and D cells can continuously power the circuit for over a year (two years for alkaline cells). The current drain is less than 0.5 mA during the battery's life. The maximum flash rate for the 3909 is 1.1 kHz; the actual flash rate of the circuit depends upon the value of the timing capacitor (300 μF in Figure 105). Decreasing this value increases the flash rate. The capacitor must be the electrolytic type. (Remember that many electrolytic devices have very broad value tolerances.)

Of course, it is also possible to build LED flasher circuits using general-purpose ICs. Figure 106 shows such a circuit built using

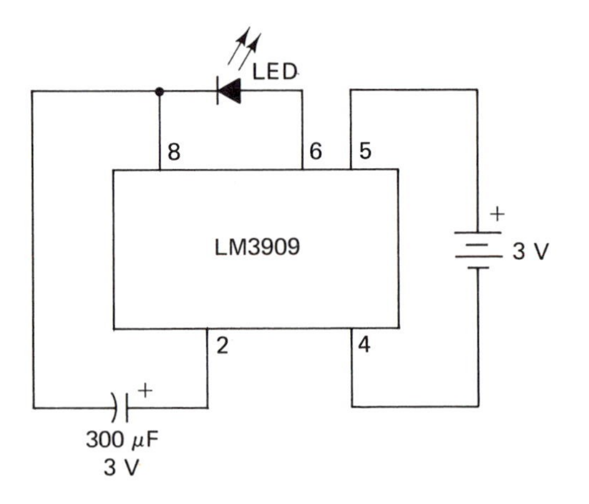

Figure 105 1 Hz flasher

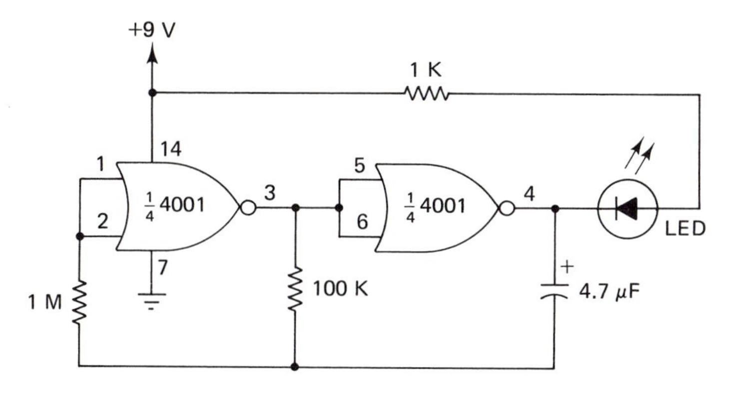

Figure 106 LED flasher

two NOR gates from a 4001 CMOS IC. As configured, the circuit will flash the LED at approximately 1–2 Hz. This rate may be altered by using a different value for the 4.7 μF electrolytic capacitor.

LED outputs for integrated circuits

Many of the circuits in this book have LEDs as their outputs. It is also possible to add LED outputs to most other IC devices, as long as the proper current-limiting resistor is used with the LED.

The first illustration in Figure 107 shows a general-purpose configuration that can be used with most linear ICs. The value of the current-limiting resistor R is given by the formula

$$R = \frac{\text{input voltage} - \text{LED voltage}}{\text{LED current}}$$

In this formula, "LED current" refers to the LED's rated forward current and "LED voltage" refers to the LED's rated forward voltage. A margin of error should be incorporated into the final value

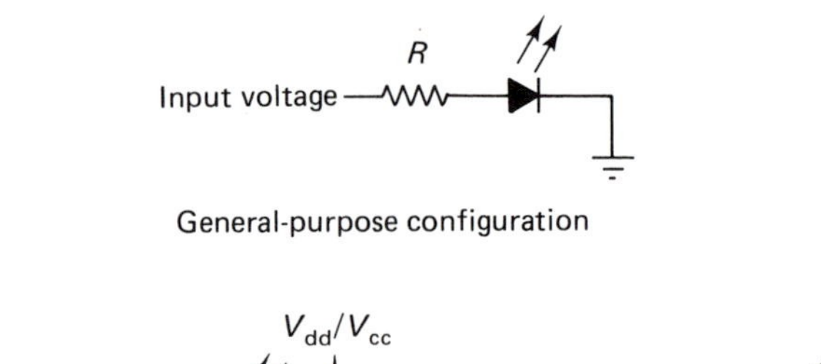

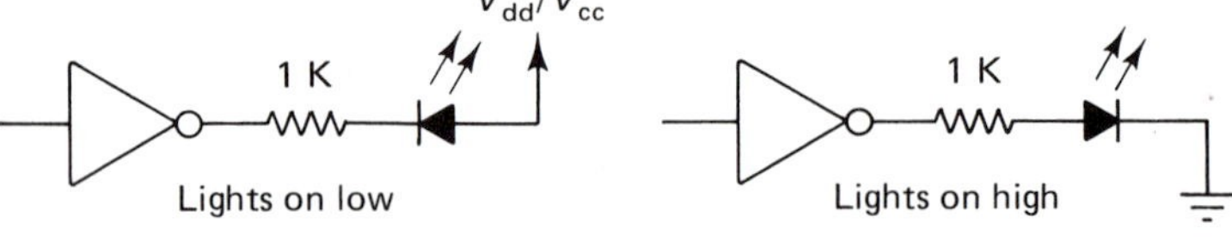

Figure 107 LED outputs

of the current-limiting resistor, since even slightly exceeding the LED's rated current capacity can destroy the LED.

The second illustration shows typical connections for CMOS and TTL devices to allow LEDs to be lit on low and high logic-level outputs. The current-limiting resistor, with value 1 K, will provide an adequate margin of safety under almost all operating conditions. When the LED is configured to light on a low logic output, the value of *V*dd or *V*cc must be identical to that used by the IC supplying the low output for the LED.

Logic gates

There is a wide variety of semiconductor logic devices available. Among these are devices containing one or more logic gates, such as AND, OR, and NAND, on a single chip with gates having two or more inputs. IC logic comes in two main families and their variations, TTL and CMOS. Although the applications and operating parameters of TTL and CMOS are quite different from each other, the principles of how logic gates can be combined to produce desired functions are the same for both.

The "universal" logic gate is the NAND. It is so called because any other type of logic gate or function can be implemented using a network of NAND gates. (In fact, even complex computer systems can be reduced to nothing more than networks of NAND gates.) The logic of a NAND gate is simple. Its output is normally high, unless all of its inputs are high, in which case the output goes to low. Popular NAND-gate ICs, such as the TTL 7400 and CMOS 4011, contain four independent two-input NAND gates on a single IC. The configuration is known as a "quad" NAND IC. Of course, a NAND gate may have more than two inputs.

Figure 108 shows some simple applications of NAND gates. The *control gate* is also known as an enabled gate. Its function is that of a standard NAND gate: both the input and control signals must be high to produce a low output. If the control input is not high, the input signal is ignored and the output remains high. The *inverter* "ties together" both inputs to make a single-input NAND device. The output of this circuit is the opposite of its input: a high input produces a low output, while a low input produces a high output.

The AND gate is a normal NAND gate followed by a second one configured as an inverter. The action of an AND gate is the

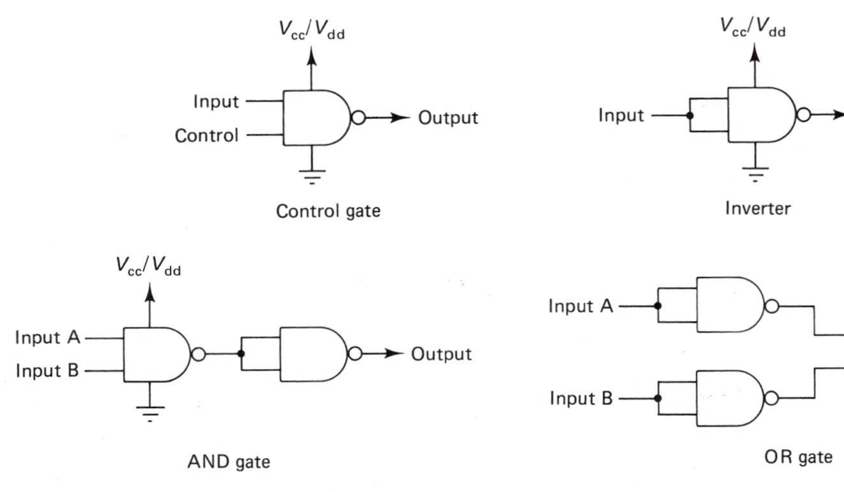

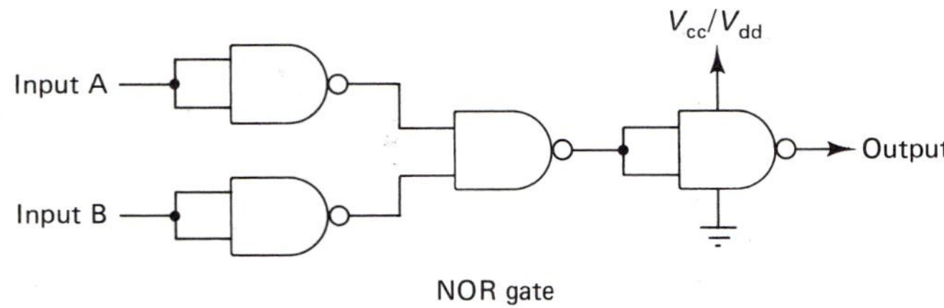

Figure 108 Logic gates

opposite of that of a NAND gate: the output of the gate is low until all inputs are high, at which time the output goes high. The OR gate is two NAND inverters whose outputs go to an ordinary NAND gate. The output of an OR gate is low only if all inputs are low; the output is high if one or more inputs are high. The NOR gate operates in opposite manner to the OR gate, producing a high input only when all inputs are low. If one or more of the inputs is high, the output is low.

Additional logic functions are shown in Figure 109. The *exclusive OR* (XOR) gate produces a high output when the input signals

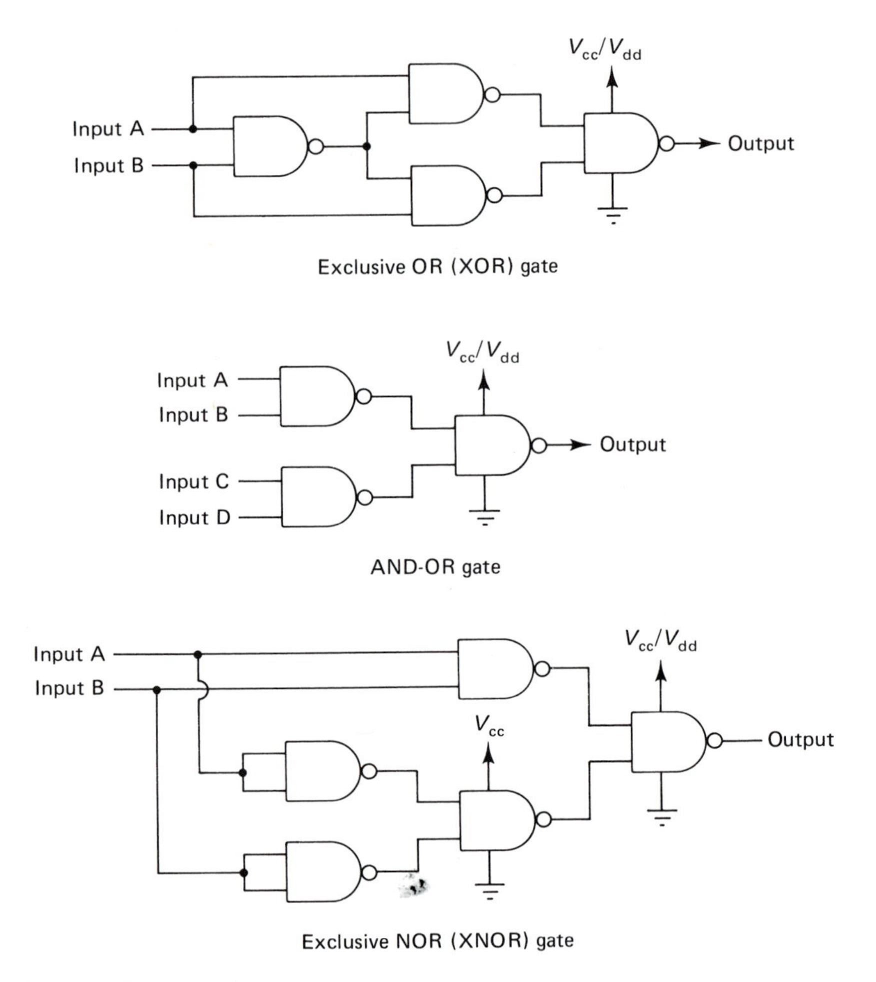

Figure 109 Logic gates

differ, and a low output when the inputs are the same (both high or both low). The AND-OR gate produces a high output when all inputs are high or when one *pair* of inputs (A and B, or C and D) is high. Otherwise, it produces a low output. Thus, if A and B are high, the output is high; but if only A and C are high (or A and D, or B and C, or B and D), the output will be low. The *exclusive NOR* (XNOR) gate operates in the opposite manner from an XOR gate: if the inputs are identical (both low or both high), the output is high; if the input signals differ, the output is low.

It is possible to implement logic functions using operational

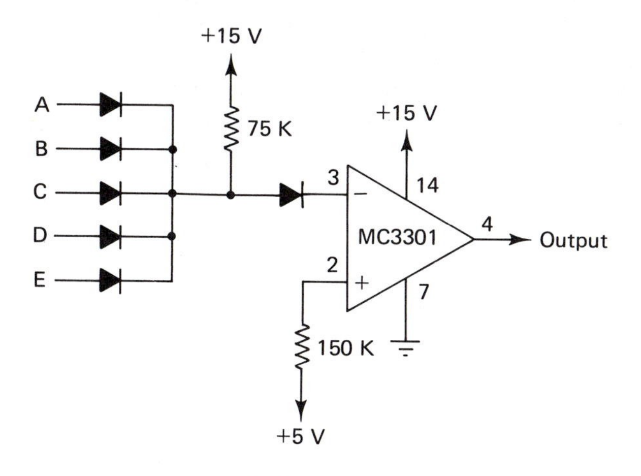

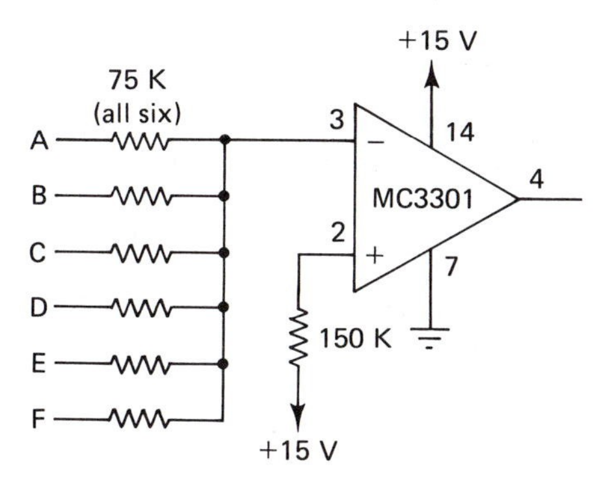

Figure 110 Logic gates

amplifiers. This is not surprising, since operational amplifier circuits were originally developed for use in analog computers. (These early op amp circuits were built from vacuum tubes.) While op amps cannot perform logic functions as rapidly or as well as digital logic ICs, they are useful for situations where it is either not desired, not practical, or not possible to properly interface digital logic with a larger analog circuit. Figure 110 shows NAND and NOR gates constructed using one section of an MC3301 quad single-supply operational amplifier device. (An LM2900 may be substituted in its place.) The diodes used by the NAND gate may be almost any general-purpose switching diodes. The supply voltage used is +15 V, and the inputs should be less than or equal to that value. As with digital devices, 0 V should be used to represent a low logic level. While the logic functions of these op amp NAND and NOR gates are identical to those of their TTL or CMOS equivalents, they do operate considerably slower and are thus not suitable for applications where rapid switching between logic states is required.

Logic probe

A logic probe is an instrument that indicates the logical status of a point in a digital circuit. Such instruments usually take their input through a narrow, sharply pointed probe and indicate the logical status of the point using LEDs. Reading such LEDs while making sure the probe is making contact at the desired point can be difficult. The circuit in Figure 111 uses a seven-segment LED display for easy reading. If the logic level is high, the LED display will read "1." If the logic level is low, the display will show "0." If the input is rapid switching between logic levels, or pulsing, the display will show "P." Each of the inverters is one section of a 7404 hex inverter IC.

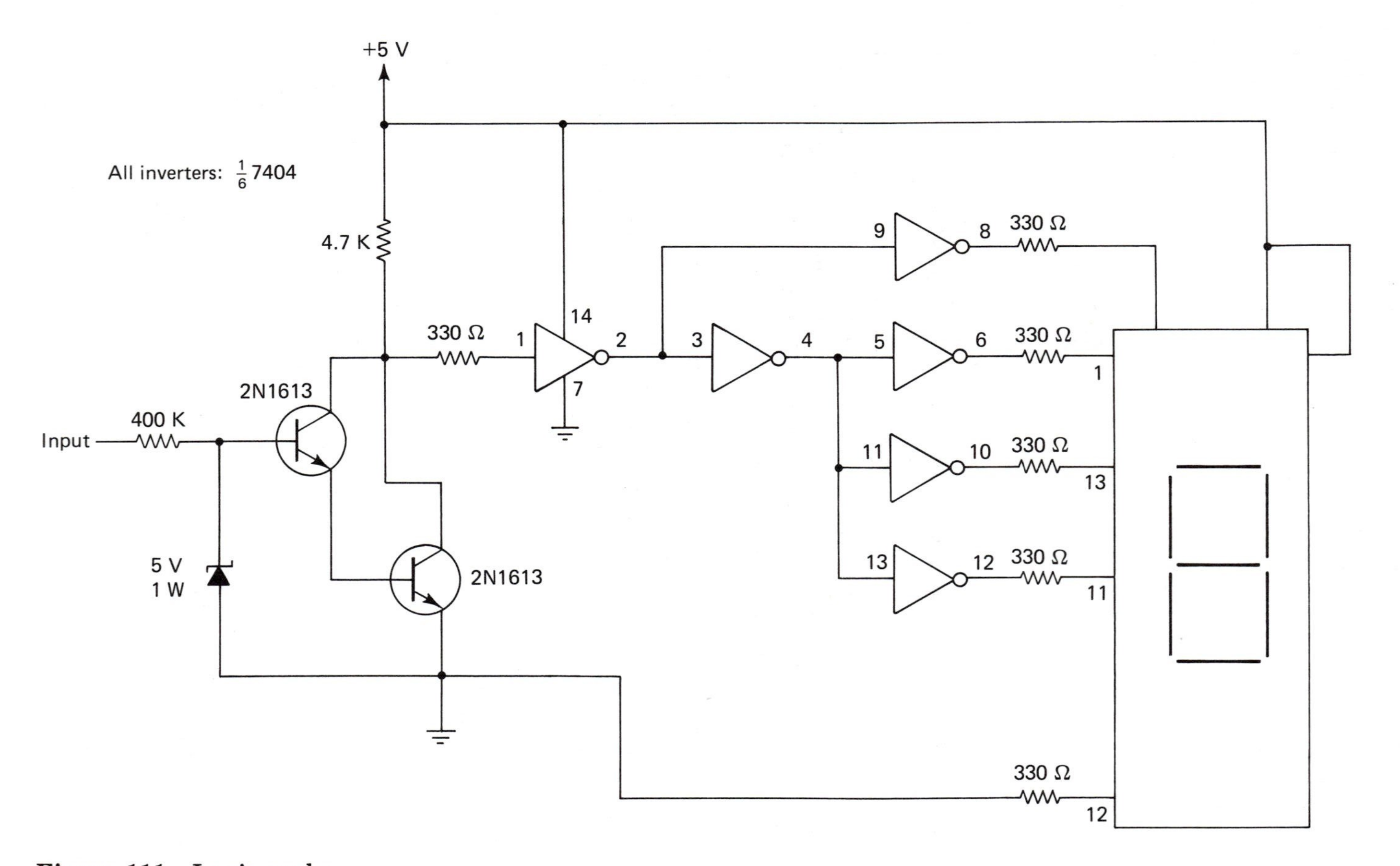

Figure 111 Logic probe

m

Mixer, audio

The circuit in Figure 112 uses one-half of an LF353 JFET-input dual-operational amplifier for high-input impedance. The resulting audio output signal is a combination of the four input signals. The gain afforded by the LF353 prevents any input signal loss, which can be a problem in passive mixer designs, and assures that each input is reflected in the resulting output according to its level relative to the other inputs. Additional inputs may be added if desired. The input signal levels should be approximately those obtained from sources such as microphone preamplifiers. Signals from audio amplifiers using the TL084 op amp IC will also work well with this circuit.

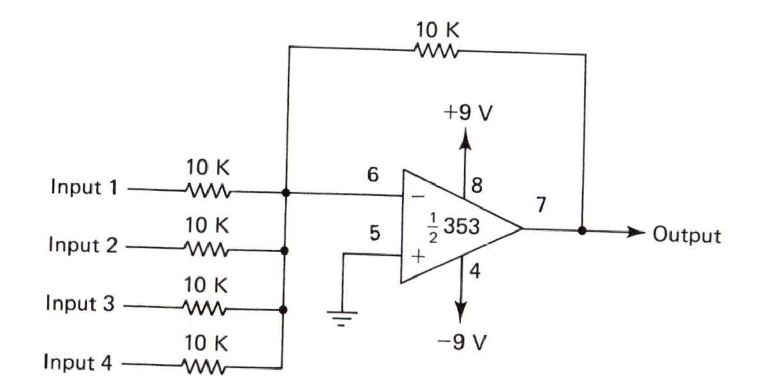

Figure 112 Audio mixer

Mixers, digital

The audio mixer circuit can work with a variety of different analog waveforms. However, it will not function well with digital (square wave) inputs. Nonetheless, it is possible to implement a digital mixer using CMOS or TTL logic. In that case, the outputs of the mixers will be the *difference* between two input signals rather than their sum.

Figure 113 shows a CMOS mixer that uses one section of a 4013 dual D flip-flop IC. If the input signals are identical in frequency, the output will be a constant high or low signal depending upon whether input 1 was high or low at positive edge of input 2. If the frequencies at input 1 and input 2 differ, the resulting output will be equal to the frequency of input 2 minus the frequency of input 1.

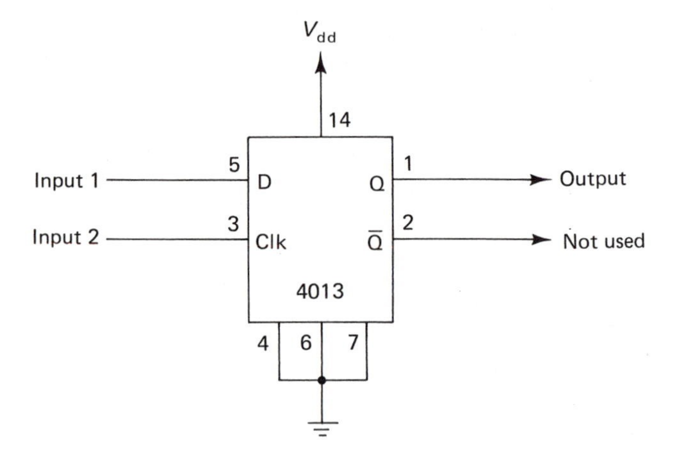

Figure 113 CMOS digital signal mixer

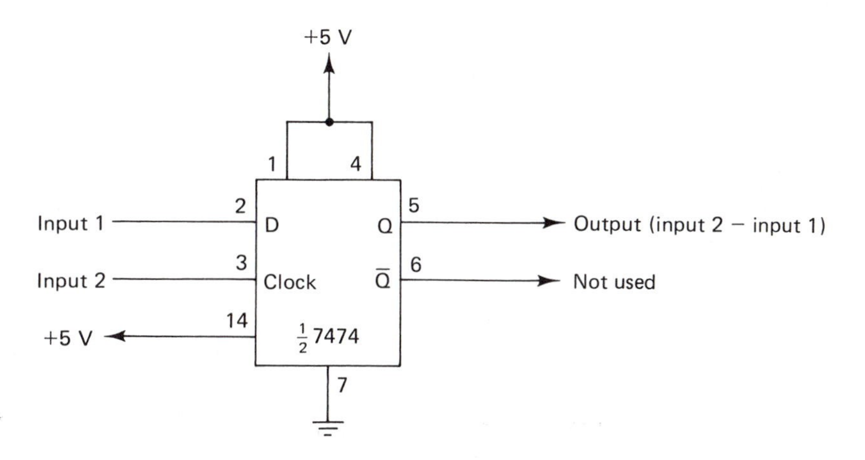

Figure 114 TTL digital signal mixer

A similar circuit is shown in Figure 114, which is implemented using one section of a 7474 TTL dual D flip-flop IC. This circuit functions in a manner very similar to the way the CMOS version functions: if the input signal frequencies are in phase, the output of the circuit will be high; if the signals differ, the output is equal to the frequency of input 2 minus the frequency of input 1.

In addition to mixing inputs, both of these circuits can be used as digital phase-difference detectors, since any change in the constant output will indicate that the two input signals are not equal in frequency. LEDs can be added to the outputs of these circuits to give a visual indication of the resulting output. Some "jitter" in the output is normal.

Modulators, amplitude

Amplitude modulation is the process by which the amplitude of the waveform making up the output of a transmitter is made to vary in accordance with the amplitude variations of a modulating waveform. Typically, this involves adding a modulating frequency to an unmodulated *carrier* waveform. The result is a new signal consisting of the unmodulated carrier frequency and a pair of *sidebands,* equal in bandwidth to the modulating frequency, located above and below the carrier frequency. For example, if a 3 MHz carrier were amplitude modulated by a 10 kHz audio signal, the result would be a carrier frequency of 3 MHz with a *lower sideband* from 2.990 MHz to 3 MHz and an *upper sideband* from 3 MHz to 3.010 MHz. Approximately two-thirds of the available transmitter power will be contained by the carrier, while the remaining one-third will be shared by the two sidebands.

One very simple approach to amplitude modulation is shown in Figure 115, which uses half of a CA3280 dual-variable op amp IC. As in the example in the previous paragraph, a carrier frequency of 3 MHz and a modulating signal of 10 kHz are used. The carrier input level is 50 mV peak to peak, while the modulation input level is 30 V peak to peak. The output of this circuit must be amplified by a class A amplifier before transmitting.

Some ICs have been developed specifically for modulation and demodulation. One is the MC1496, which was developed for balanced modulation (the production of two sidebands, but with the carrier suppressed). However, it can also easily serve as an amplitude modulator, as shown in Figure 116. The precise amount of carrier remaining is controlled by the 50 K potentiometer, which should be adjusted for proper balance between the carrier and the sidebands. The operating range of this circuit is typically between 1,000 kHz to 10 MHz.

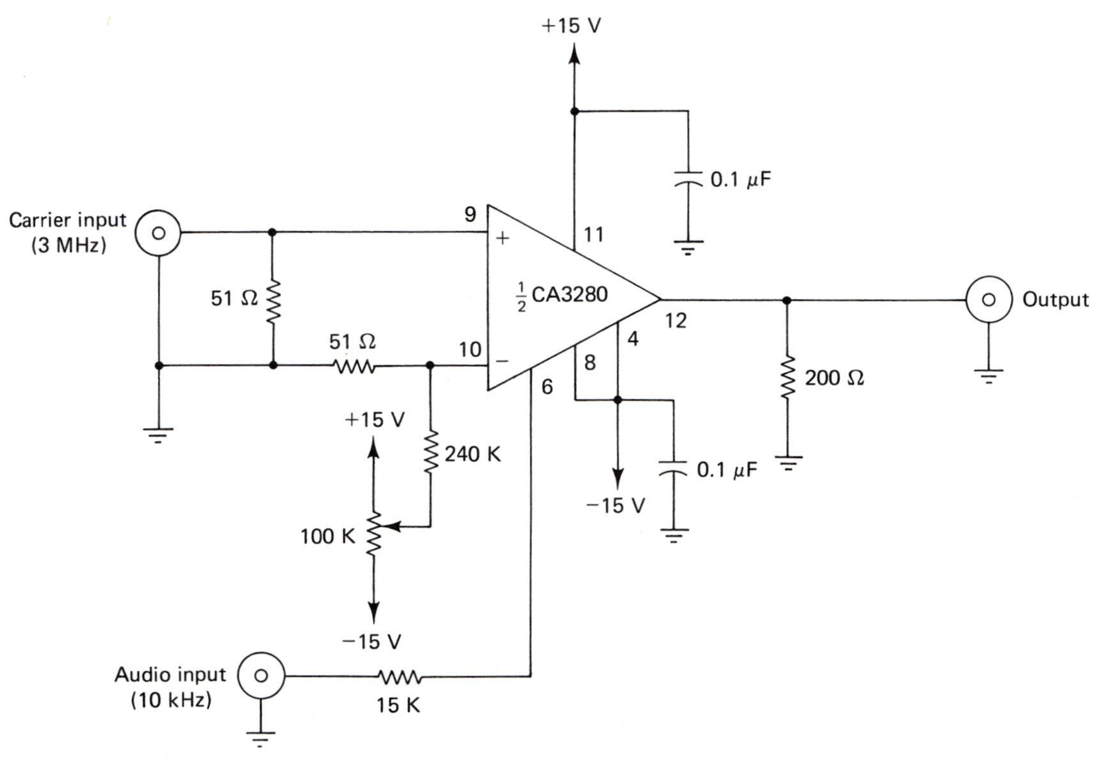

Figure 115 Amplitude modulator

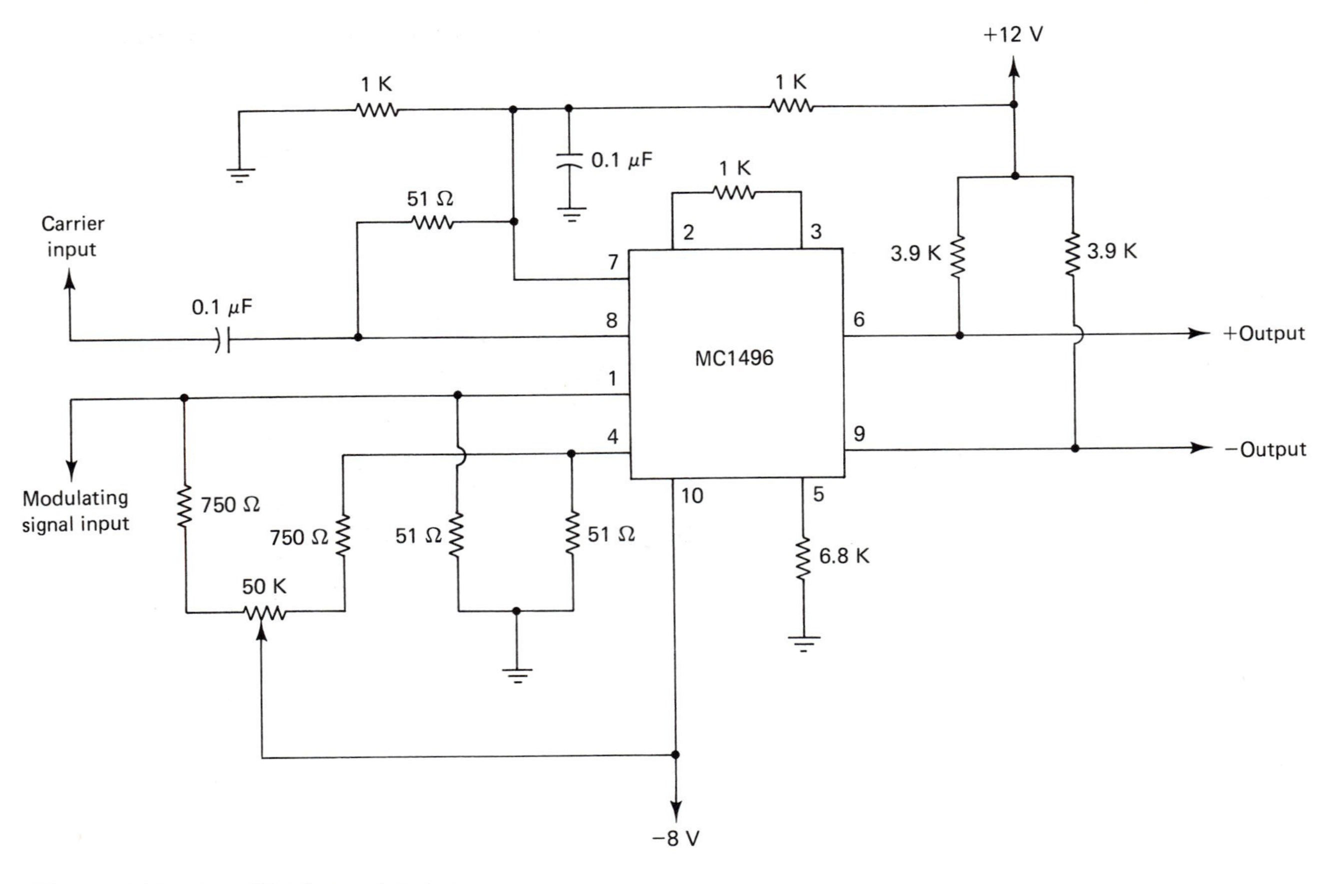

Figure 116 Amplitude modulator

Modulator, balanced

A balanced modulator is similar to an amplitude modulator, except that it produces an output consisting of two sidebands *without* a carrier. This circuit is often used in single-sideband (SSB) transmitter systems. In an SSB transmitter, one of the two identical sidebands is suppressed (usually by means of a crystal or mechanical filter which passes one sideband while rejecting the other), while the other is amplified and transmitted. Figure 117 shows a balanced modulator circuit which uses an MC1496 IC which has been specifically designed for such use.

The circuit shown is very similar to the one in Figure 117; the only significant differences are in the values of certain components. The 50 K potentiometer is adjusted until the carrier in the output is removed. The carrier suppression is excellent, with −50 dB at 10 MHz being a typical figure. Operating range is usually 500 kHz to over 10 MHz.

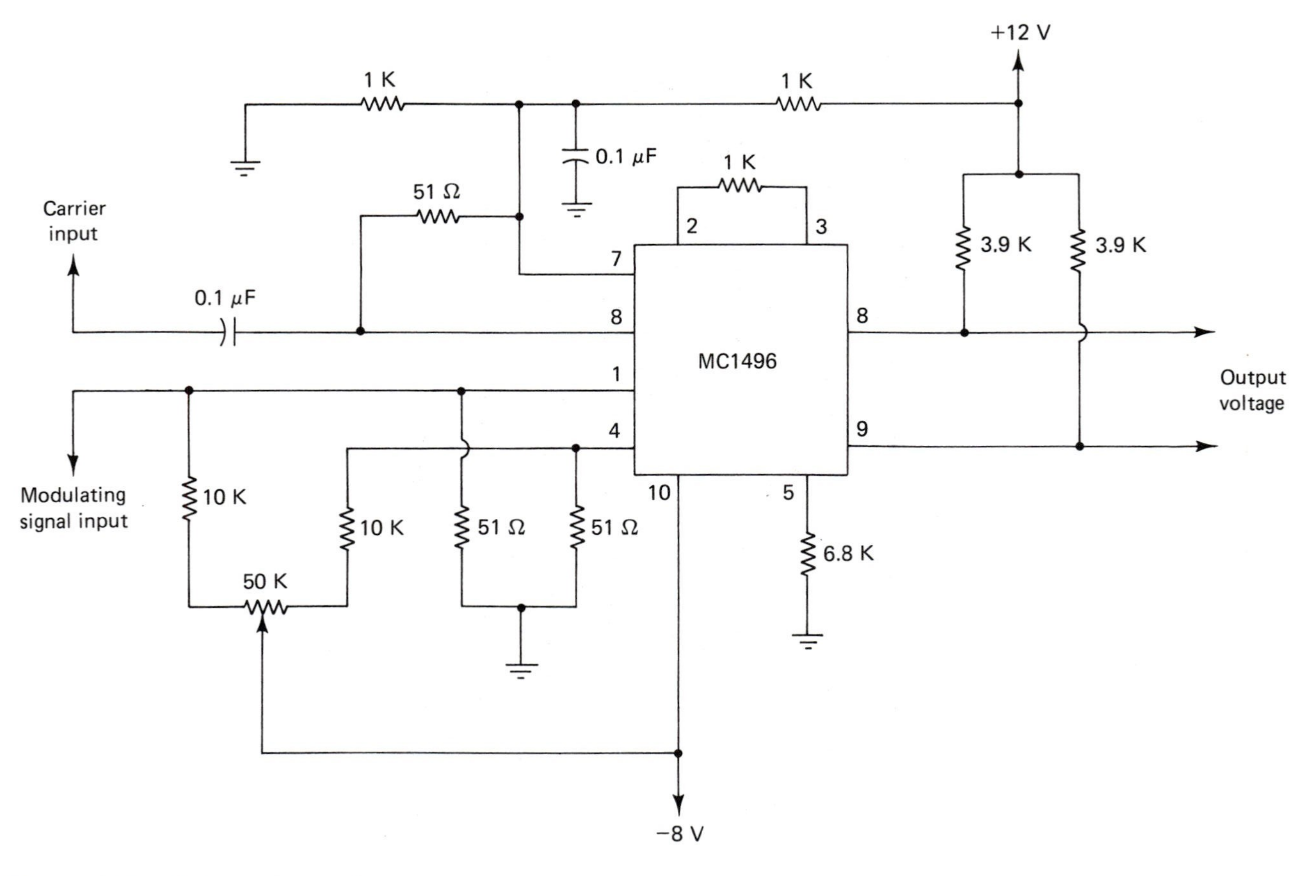

Figure 117 Balanced modulator

Modulator, pulse width

A pulse width modulator is a modulator in which the duration (or width) of the output pulses produced by the circuit varies in accordance with the modulating signal applied. In effect, it can be thought of as a square wave oscillator whose output pulses are stretched or compressed by a modulating signal. Figure 118 shows a simple pulse

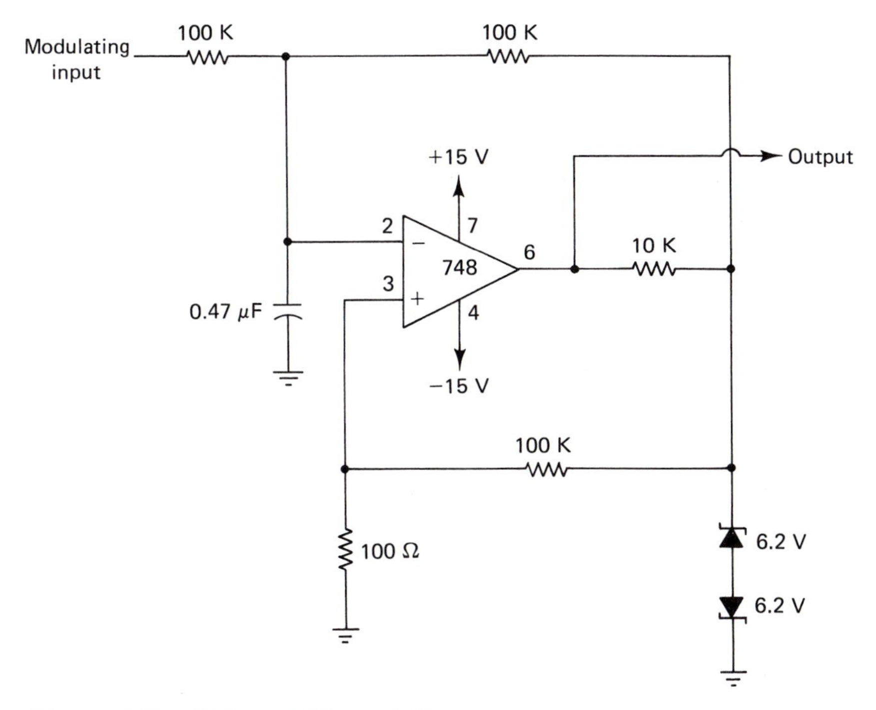

Figure 118 Pulse width modulator

width modulator that uses the 748 operational amplifier. The modulating signal can vary between −5 and +5 V, and the output pulses will vary in width according to the modulation applied. The linearity of this circuit is not great, but it is adequate for most experimental and hobbyist applications as well as for demonstrating the operation of a pulse width modulator.

Modulator, video

The development of home video game systems sparked a demand for simple but effective video modulator systems. One IC, the MC1374, was developed by Motorola for use in videogame modulator sections. The circuit in Figure 119 uses the MC1374 in a video modulator to produce an output on television channels 3 or 4, as selected by a switch. The supply voltage for this circuit can range from +5 to +12 V DC. Inductor $L1$ consists of four turns of #22 wire wound $\frac{1}{4}$ inch in diameter, while inductor $L2$ consists of 40 turns of #36 wire wound $\frac{3}{16}$ inch in diameter. Video input should be 1 V peak. Output is approximately 70 mV peak to peak, with an RMS value of about 12 mV. This is 12 dB greater than current FCC rules permit, so it must be "padded down" for commercial applications. Care must be taken in constructing this circuit to ensure proper operation, in particular in placement of parts and in the values of the inductors used.

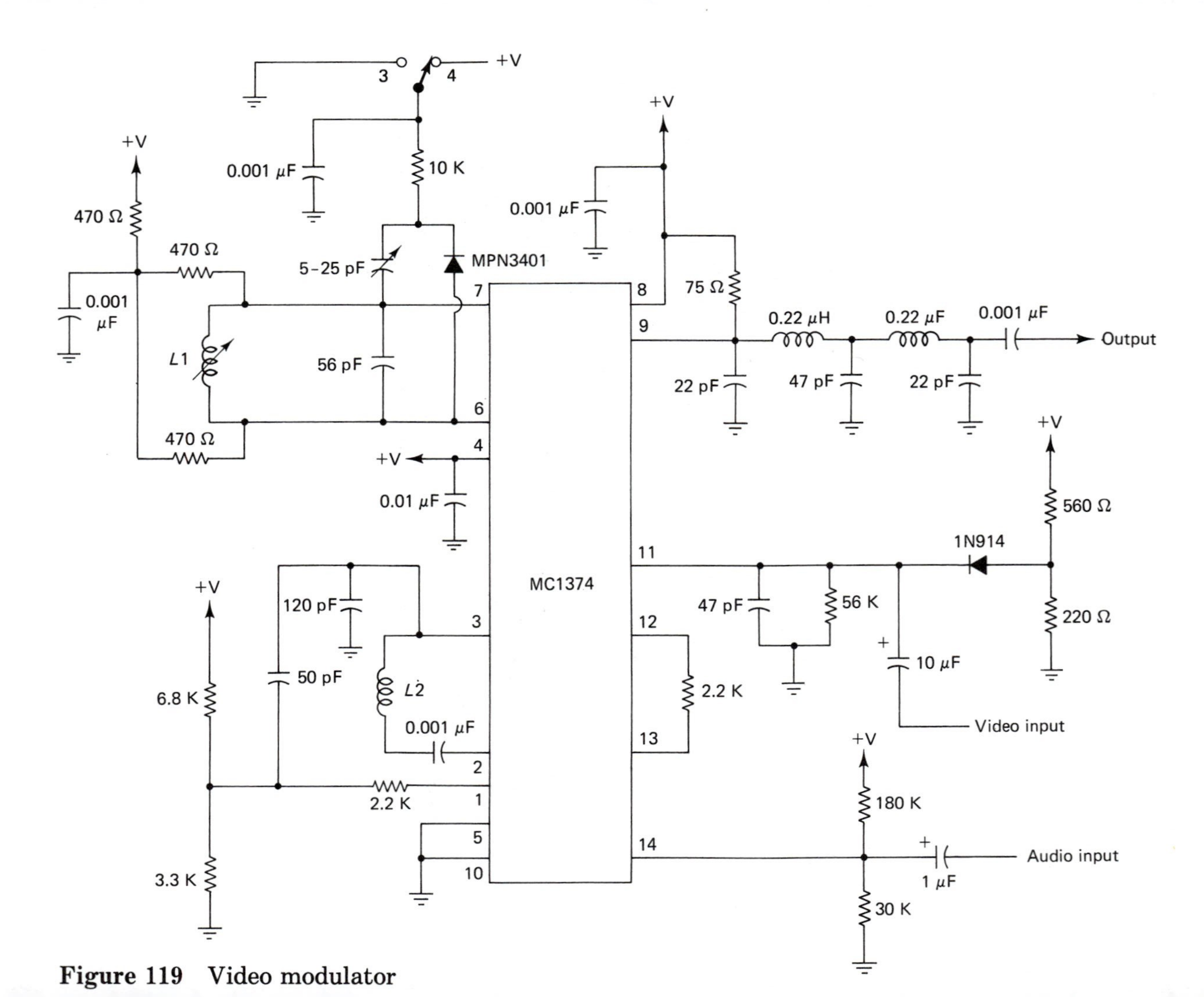

Figure 119 Video modulator

Multiplier, voltage

A voltage multiplier is a circuit which gives the linear product of two input voltages. The circuit in Figure 120 shows a highly linear voltage multiplier that uses two ICL8048 logarithmic amplifiers, one ICL8049 antilogarithmic amplifier, and one 741 operational amplifier. The two input voltages, X and Y, produce an output given by the formula

$$\text{Output voltage} = \frac{XY}{10}$$

where X and Y are DC voltages from $+0.1$ to $+10$ V. In operation, the outputs of the ICL8048 devices are summed by the 741 op amp, whose output is then used by the ICL8049 to produce an output voltage which is proportional to the product of the inputs X and Y. Accuracy is 1 percent or better at $+10$ V output levels.

A simpler but less accurate voltage multiplier based upon the MC1594 four-quadrant multiplier IC is shown in Figure 121. The values of input resistors $R1$ and $R2$ for linear operation of the circuit depend upon the anticipated values of the input voltages. The value of $R1$ is equal, in kilohms, to six times the input voltage at input Y. In similar fashion, the value of $R2$ is equal, in kilohms, to three times the input voltage at X. Suppose, then, that the input voltage at both X and Y is 3 V. In such a case, $R1$ would be 18 K while $R2$ would be 9 K. For wideband operation with several different input voltages, values of 52 K for $R1$ and 30 K for $R2$ can be used. However, accuracy and linearity are reduced by this approach. The output of this circuit is proportional to the product of the two input voltages in a manner similar to the multiplier circuit of Figure 120.

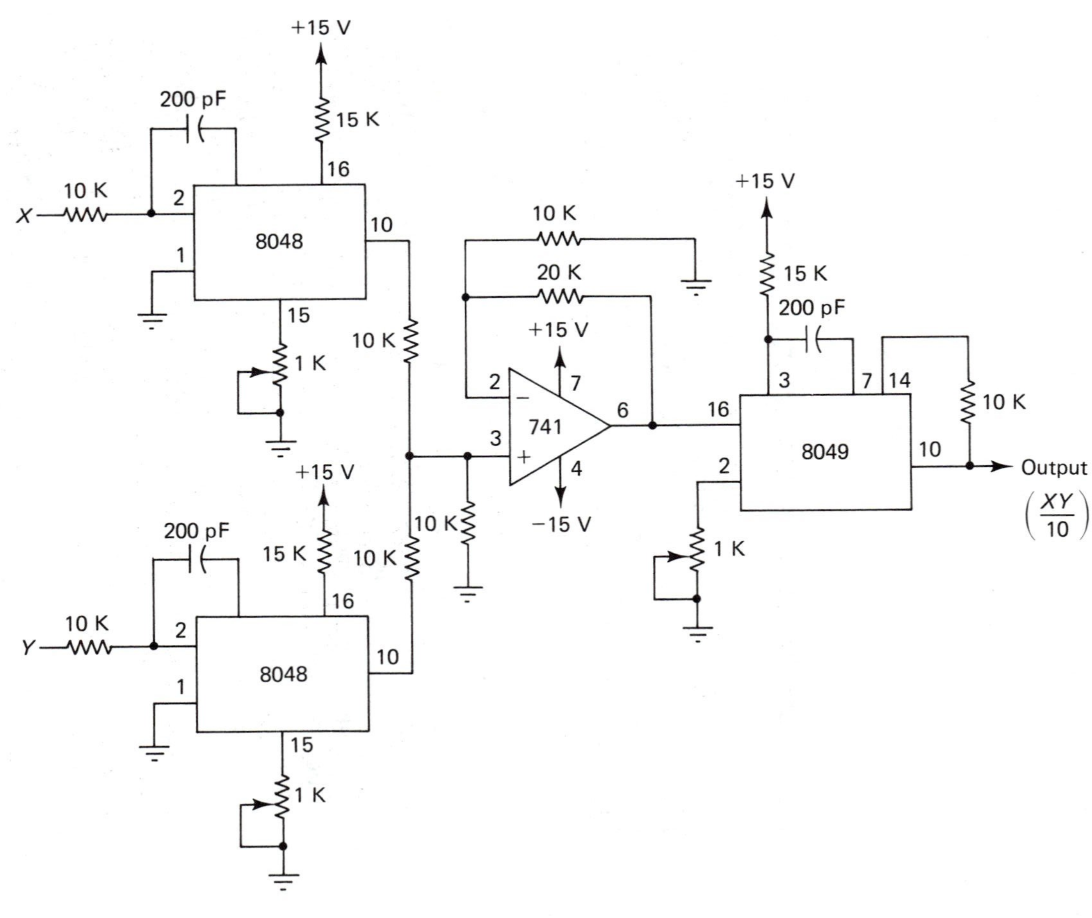

Figure 120 Linear DC voltage multiplier

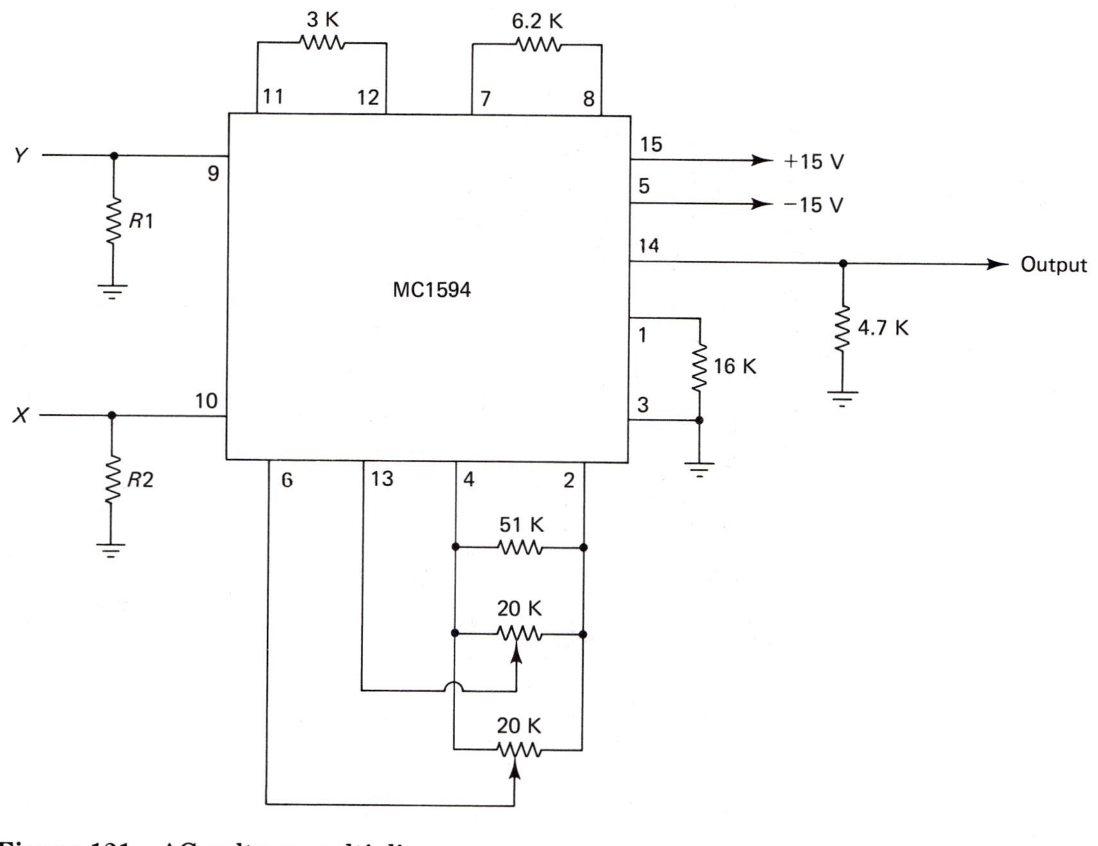

Figure 121 AC voltage multiplier

Multivibrators, astable

An astable multivibrator is also termed a *free-running* multivibrator because a circuit of this type produces a continuous stream of output pulses at a given frequency determined by the circuit's characteristics.

Perhaps the most common approach to implementing an astable multivibrator with a desired output frequency is to use a timer IC, such as the 555, with appropriate external components to determine the output frequency. Figure 122 shows a circuit using the 555 whose outputs can be used as clock or input signals for TTL

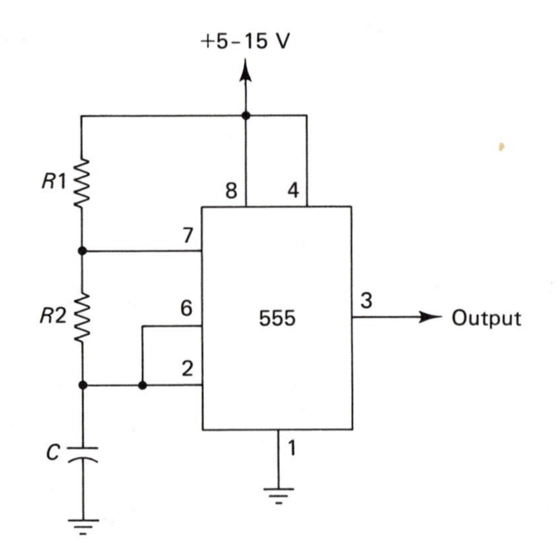

Figure 122 Astable multivibrator

ICs. The output frequency depends upon the values of $R1$, $R2$, and C according to the formula

$$\text{Output frequency} = \frac{1.44}{C(R1 + 2\,R2)}$$

where the minimum value of $R1$ or $R2$ should be at least 1 K and the minimum value of C should be at least 500 pF. Also, the total value of $R1$ plus $R2$ must not exceed 3.3 M, and it is recommended that C be 1,000 pF unless low-value resistors are used. A variable-frequency astable multivibrator can be created by using a 1 M potentiometer for $R1$, a 1 K fixed resistor for $R2$, and a 0.01 μF capacitor for C. The maximum output frequency is approximately 1 MHz, although best operation is at lower frequencies of a few hundred kHz.

While the 555 timer is usually the best way to go to create astable multivibrators, it is also possible to use other devices, such as operational amplifiers, in such circuits. Figure 123 shows a circuit using a 311 operational amplifier which supplies a constant 100 kHz square wave output. A +5 V power supply is used, which makes the output of this circuit TTL compatible. Higher voltages, such as +15 V, may be used if the outputs will not drive TTL logic.

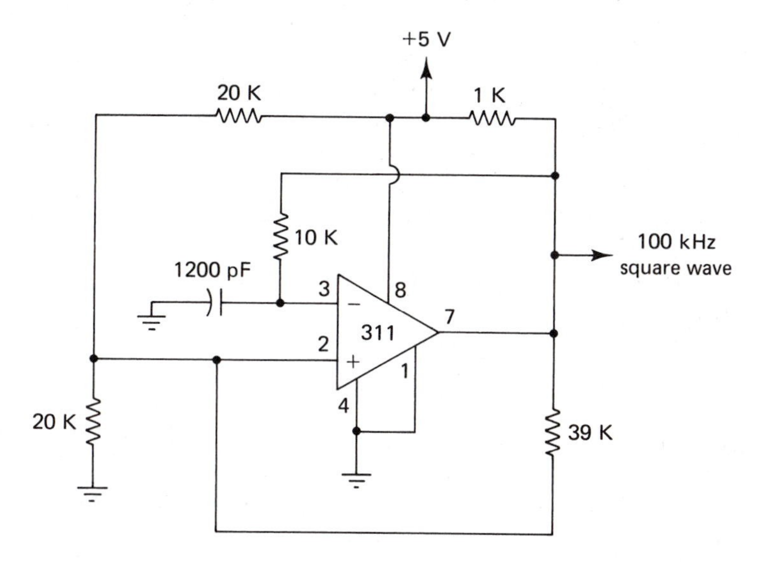

Figure 123 100 kHz astable multivibrator

Multivibrators, monostable

A monostable multivibrator is a circuit which produces a *single* output pulse rather than a stream of pulses, and is therefore often described as a *one-shot* multivibrator. The output pulse is triggered by an input signal known appropriately as the *trigger* input. Since the output pulse of a monostable multivibrator is positive (i.e., an increase from zero voltage to a desired voltage level), the trigger input usually involves switching a constant positive voltage to zero voltage. Once triggered, the output pulse begins and most monostable multivibrators will ignore further trigger input signals until the output pulse is completed. Moreover, most monostable multivibrators will need a "recovery period" after the completion of an output pulse before they can be triggered again. (Some of these operating restrictions are overcome in ICs specifically developed for use as multivibrators.) The duration of the trigger input has no relation to the length of the output pulse.

The 555 timer IC is frequently used in monostable multivibrators. Figure 124 shows such a circuit in which the length of the output pulse is controlled by the values of R and C. The formula for determining the output pulse length is

$$\text{Output pulse length} = 1.1RC$$

where R should be in the range of 1 K to 3.3 M and C should be at least 500 pF. The output of this circuit is TTL compatible if a +5 V power supply is used; otherwise, power supplies up to + 15 V may be used.

One of the monostable multivibrator ICs which has been developed is the 74123. This device contains two *retriggerable* monostable multivibrators in a single package. "Retriggerable" means that the device can accept trigger input signals at any time, and

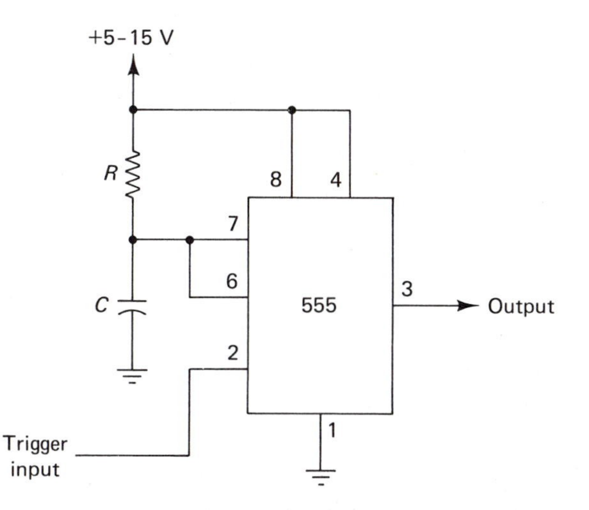

Figure 124 Monostable multivibrator

that by proper spacing of the trigger inputs the output of such a circuit can be continuously held at a high logic level. (This type of circuit is sometimes described as a negative-recovery circuit.) Another interesting aspect of the 74123 is that each multivibrator it possesses has two *complementary* outputs, Q and $\overline{Q}$, which always have opposite states. Thus, when Q goes high, $\overline{Q}$ will go low and vice versa. The 74123 is also capable of very brief output pulses. Figure 125 shows the 74123 used to provide two monostable multivibrators. The formula for determining the output pulse time is the same as for the 555 timer circuit, using each resistor-capacitor pair ($R1$ and $C1$ or $R2$ and $C2$). Each multivibrator is independent of the other; they may have different timing cycles and be triggered separately. The 74123 is a TTL device, and its power supply must be held to +5 V.

The 74123 is sensitive to "noisy" trigger signals, and proper component layout is essential for correct operation. For these reasons, the 555 is preferred in most monostable multivibrator circuits unless extremely short output pulses or complementary outputs are needed.

Sometimes a monostable multivibrator with a very long output period is needed, and these can be more easily implemented using op amps rather than such ICs as the 555. Figure 126 shows an MC34001 JFET input operational amplifier configured to act as a monostable multivibrator; with the parts values indicated in parentheses, a 100-second period is produced. Different periods can be

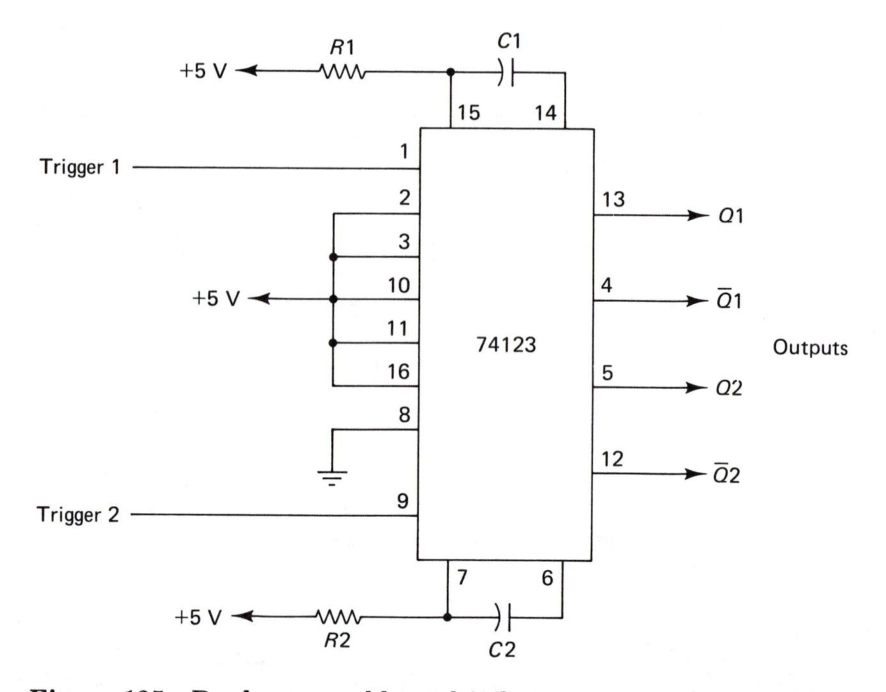

Figure 125 Dual monostable multivibrator

produced using different component values. However, in all cases, $R1$ must equal $R2$, $R3$ must equal $R4$, $R6$ should be 10 times the value of $R5$, and capacitor C should be of the polycarbonate or polystyrene variety. The switch controls operation of this circuit. When set to "run," the multivibrator is triggered. When set to "clear," the circuit is reset.

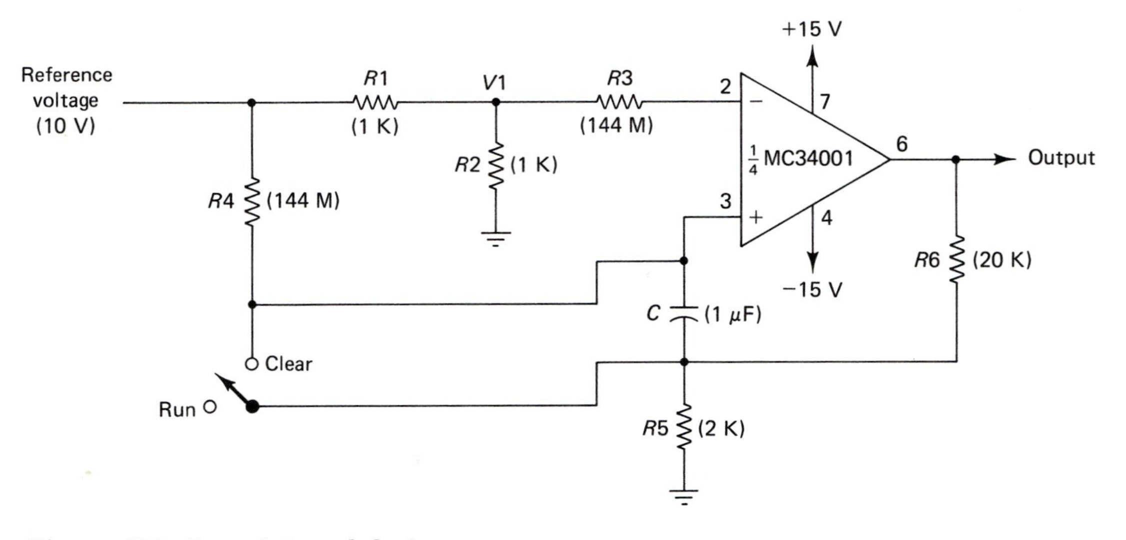

Figure 126 Long-interval design

n

Noise sources

Pink noise is noise whose amplitude is inversely proportional to its frequency over a certain range. Most people find pink noise pleasant to listen to; surf and raindrops are two close approximations of pink noise in nature. By contrast, *white noise* is a random noise in which frequency and amplitude are not related; most people find it disconcerting or unpleasant. Both types of noise are useful in testing audio circuits and devices such as speakers and headphones, and some projects require a reliable source of either type of noise.

Both pink and white noise can be produced by using special configurations of audio oscillators, but a more convenient approach is to use the MM5837 (also known as the S2688), an IC specifically developed to generate both types of noise. Figure 127 shows how this device is used. In both circuits shown, the output must be applied to an audio amplifier before it can be used. The value of the resistor used in the pink noise generator may range from 1 K to 10 K, while the capacitor may range from 0.01 μF to 0.2 μF; changing these values will alter the frequency spectrum of the pink noise output. The frequency spectrum can also be altered by reducing the supply voltage.

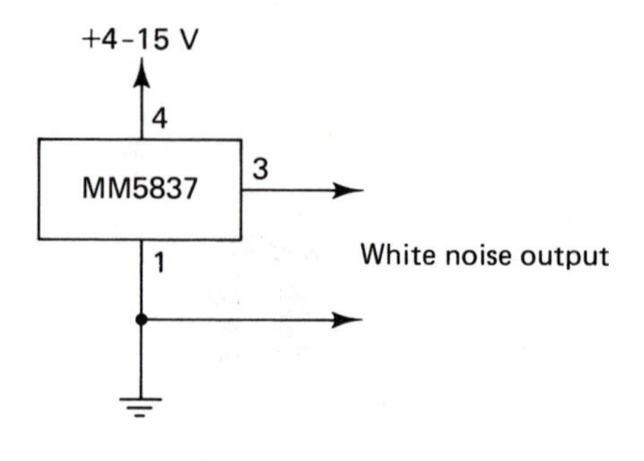

White Noise Source

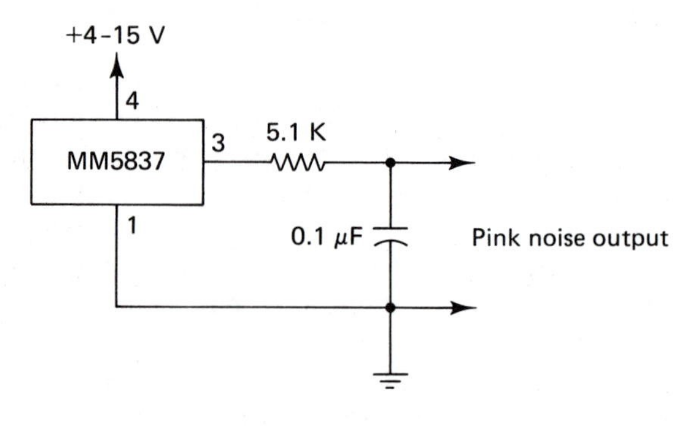

Pink Noise Generator

Figure 127 Noise production

Notch filter

A notch filter works by passing a range of frequencies, except for a narrow segment, or "notch," which it rejects. A good notch filter should have a very sharp rejection curve, meaning that attenuation should be minimal until the undesired frequency range is reached. Frequencies above and below the notch are passed without attenuation. There are several approaches to implementing notch filters, but using a high-performance operational amplifier device is usually the easiest method. Figure 128 shows a notch filter that uses the MC33171 op amp, a recently introduced high-performance device which has a wide bandwidth and high slew rate. With the parts values shown, the notch frequency will be 1 kHz. Other notch fre-

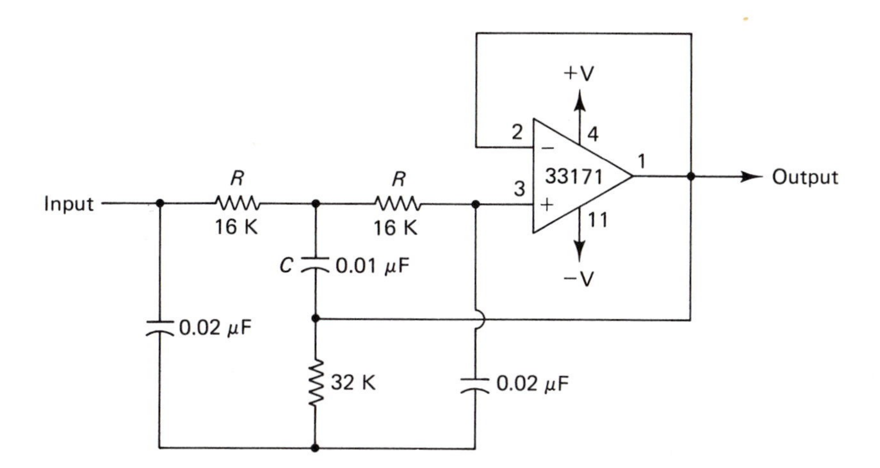

Figure 128 Notch filter

quencies can be obtained by varying the values of the two resistors labeled R and the capacitor C according to the formula

$$\text{Notch frequency} = \frac{1}{4\pi RC}$$

(Both resistors must have the same value.) The dual-polarity supply voltage may range from 1.5 to 22 V DC, and the input signal voltage must be 0.2 V or greater.

O

Oscillator, alternating tone

The output of this audio oscillator, shown in Figure 129, is an alternating two-tone sound similar to that produced by police and ambulance sirens in Europe and Japan. The MC3524 device used is a dual-voltage comparator IC intended for use in power supply supervisory circuits. By firing the comparators in alternating sequence, the distinctive two-tone sound is produced. The output of the circuit will drive most small 8 Ω speakers.

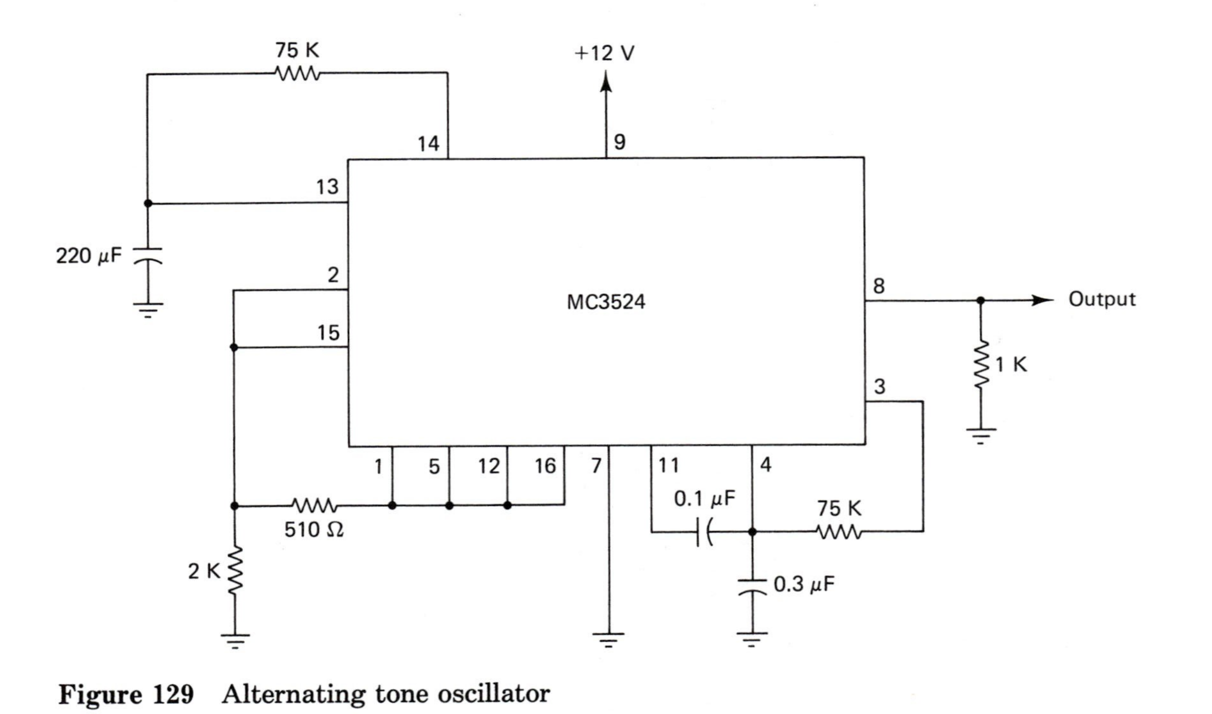

Figure 129 Alternating tone oscillator

Oscillator, audio

Audio oscillators can be built in numerous ways using a variety of ICs or discrete components. However, generating different waveforms with high accuracy from a single circuit can be a problem. One solution is the ICL8038 precision waveform generator device, which can produce sine, square, triangular, sawtooth, and pulse waveforms. Figure 130 shows this IC used in a circuit to produce simultaneous sine, square, and triangular waveform outputs at audio frequencies ranging from 20 Hz to 20 kHz. The output frequency is determined by the 100 K potentiometer connected to pin 8, while the other 100 K potentiometer connected to pin 12 adjusts the sine wave output for lowest distortion. The 10 M potentiometer and the 1 K potentiometer set the duty cycle for the oscillator. When properly adjusted, the sine wave output can have 1 percent distortion or less, while the triangular waveform can be linear to within 0.1 percent. The triangular and sine wave outputs will be in phase with each other. Note that the ICL8038 will often run "hot" and will feel very warm to the touch during operation—this is normal.

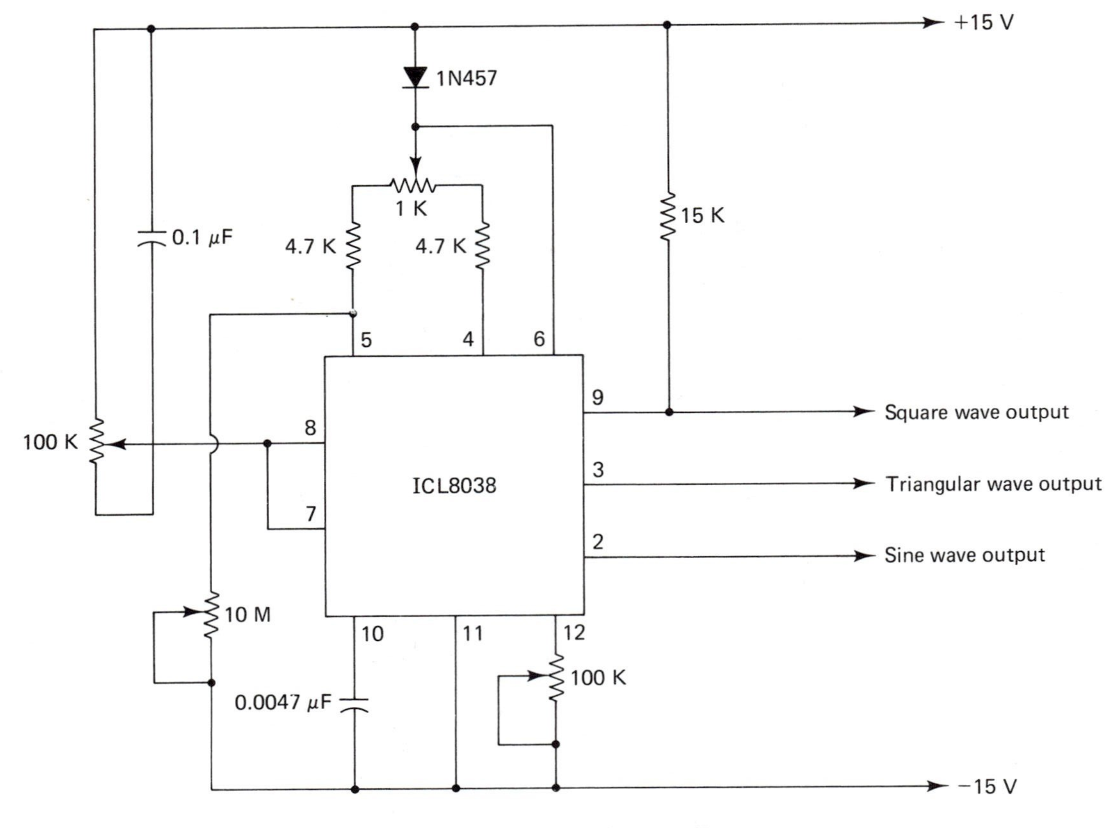

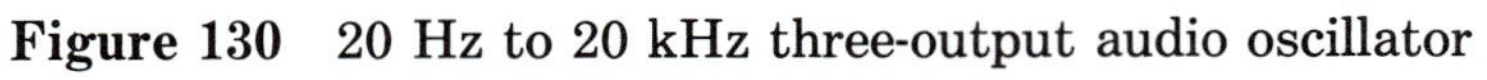

Figure 130 20 Hz to 20 kHz three-output audio oscillator

Oscillator, Colpitts

The Colpitts oscillator operates at RF frequencies and has a feedback loop consisting of a parallel LC (inductance and capacitance) circuit to make the circuit oscillate. A distinguishing feature of the Colpitts configuration is that the capacitance in the LC circuit is

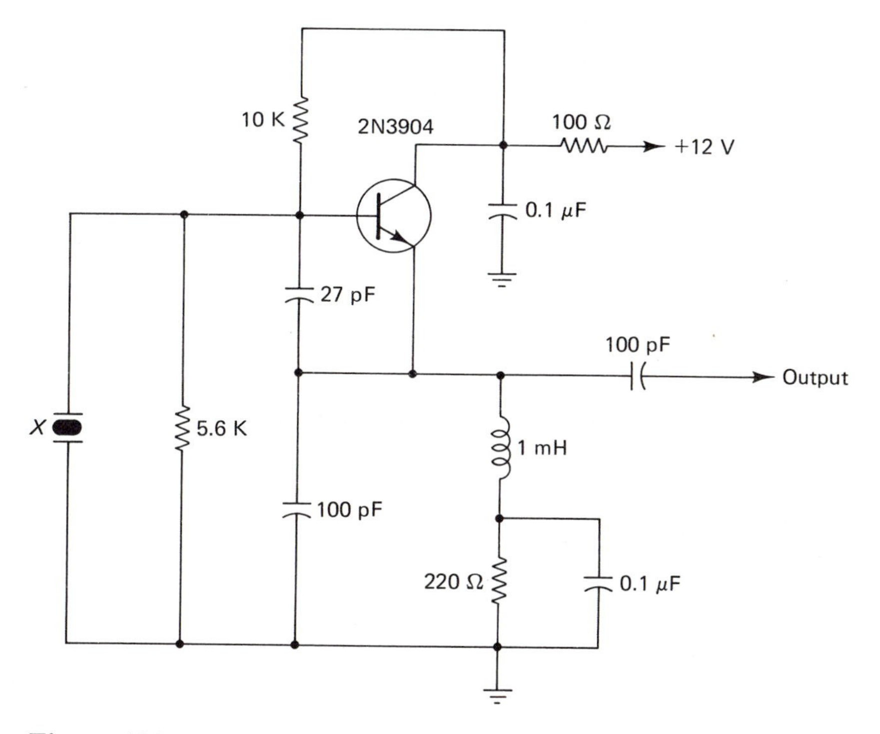

Figure 131 2.5–5 MHz Colpitts oscillator

formed from two capacitors connected in series and then split to form a voltage divider. Figure 131 shows a single-transistor Colpitts circuit capable of producing an unmodulated carrier on frequencies between 2.5 to 5 MHz. The operating frequency of the oscillator is determined by X, which should be a fundamental-mode crystal "cut" to the desired operating frequency. The output of this circuit is a few milliwatts. As with all RF circuits, placement of components and good construction practice are important, and "debugging" the final circuit may take some patience. This circuit must not be connected to a radiating antenna unless the appropriate operator and station licenses are held. If the power supply is interrupted by a telegraph key, the circuit makes an excellent low-power transmitter for the 3,500–4,000 kHz amateur radio band.

Oscillator, digitally tuned audio

Most variable-frequency audio oscillators are tuned to different frequencies using potentiometers or, in a few designs, variable capacitors. Tuning of such oscillators from digital circuitry can be a problem. However, a few ICs have been developed which can be configured as digitally tuned audio oscillators. One of these is the S3528 programmable low-pass filter, shown in Figure 132 as an audio oscillator capable of operating from 40 Hz to 20 kHz in 64 discrete steps in accordance with a six-bit control word at inputs D0 through D5. If all inputs are at a low logic level, the output is 40 Hz; if all are high, the output is 20 kHz. The internal clock of the S3528 is controlled by a 3.58 MHz "color burst" crystal, labeled X. Pin 4 is the clear input; if it is at a high logic level, the inputs are loaded and further changes at the inputs have no effect on the oscillator's output. To load new inputs, return pin 4 to ground momentarily, and then return it to a high logic level. For proper operation, this circuit requires both digital and analog ground points. If desired, switches may be used at the inputs to select between high or low logic levels, or the outputs of CMOS devices may be used.

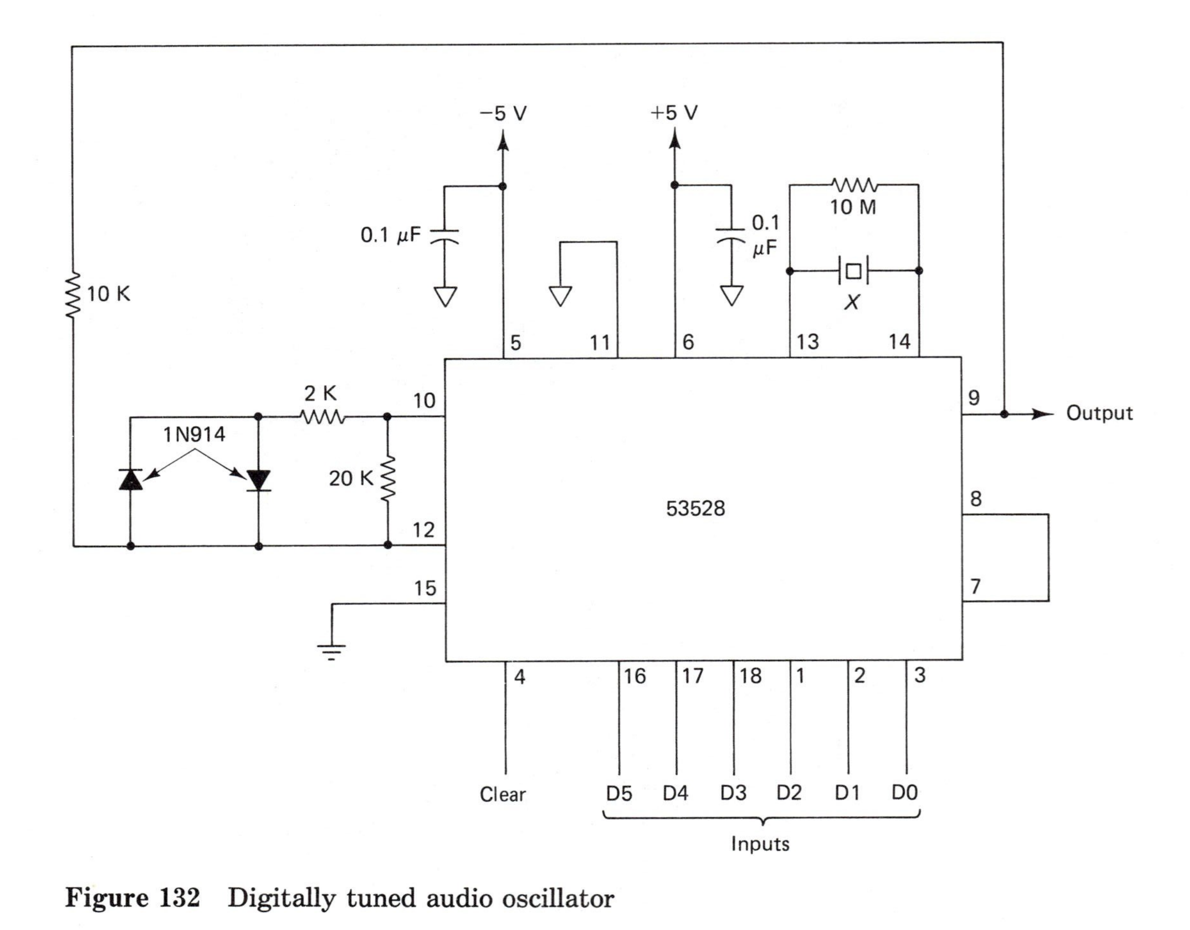

Figure 132 Digitally tuned audio oscillator

Oscillator, dual-frequency

Square wave oscillators whose output frequencies can be varied are common enough, but it is much more difficult to obtain two square wave outputs of different frequency simultaneously from the same oscillator circuit. One interesting approach is shown in Figure 133, which is a circuit that can produce outputs of approximately 1.5 and 4 kHz from a single IC. The device in this case is a 567-tone decoder. The original purpose of this IC was in tone-dialing telephone systems, where it was used to recognize the dialing tones

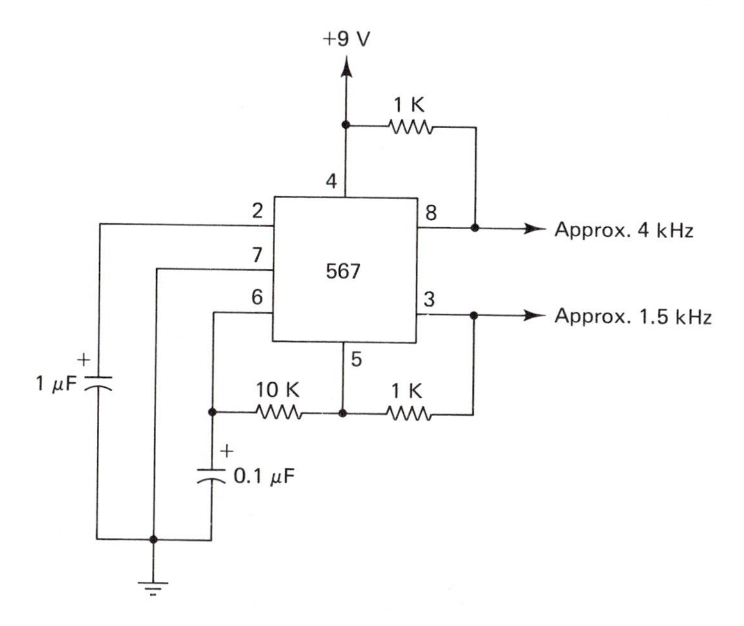

Figure 133 Dual-frequency oscillator

received. The heart of the 567 is a phase-locked loop for high stability and accuracy. In this application, the decoder input (pin 3) is actually used as an output. The actual output frequencies will vary somewhat from those indicated depending upon the precision of the resistors and capacitors used.

Oscillator, dual-phase

The dual-frequency oscillator is not the only interesting oscillator which can be implemented using the 567-tone decoder IC. The 567 can also be used to produce a circuit with dual square wave outputs having the same frequency, but out of phase with each other. The circuit in Figure 134 produces two 1 kHz square wave outputs, but the signal at output 1 leads the signal at output 2 by approximately half a square wave cycle: the signal at output 2 drops to zero during

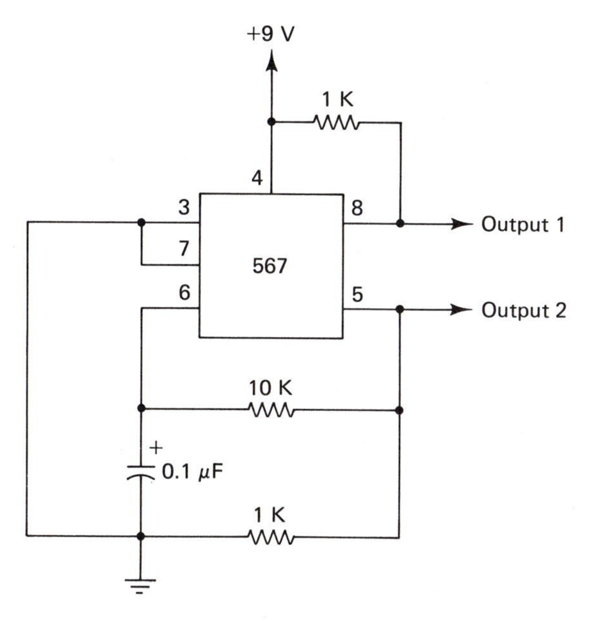

Figure 134 Dual-phase oscillator

mid-duration of the square wave peak of output 1. Both outputs reach a maximum value of 5 V. This circuit can provide two logic clock signals of the same frequency, but with slightly different phases.

Oscillator, Morse code practice

The 555 timer IC can easily be configured as a simple oscillator for generating tones used for practicing the international Morse code. The circuit in Figure 135 shows this "bare bones" oscillator in which a telegraph key is in series with the positive supply voltage; the keying action turns the oscillator on and off to form the Morse code characters. The output of the 555 drives an 8 Ω speaker. The key may be replaced with any other type of switch or other device to control the flow of +9 V to the circuit; this allows the circuit to be used as an alarm or siren.

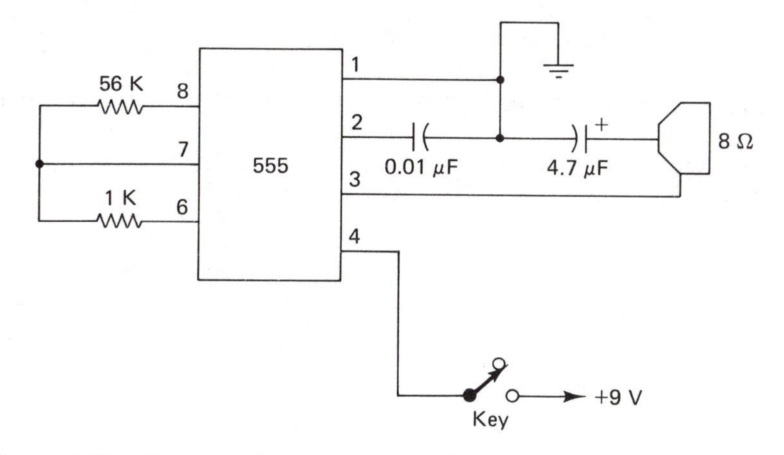

Figure 135 Morse code practice oscillator

Oscillator, Pierce

A Pierce oscillator, shown in Figure 136, is basically a variation of the Colpitts oscillator in which the frequency-determining element is located between the base and collector of the transistor. The unmodulated carrier output power of this circuit is a few milliwatts, and the output frequency is determined by X, a fundamental-mode-type crystal. The operating frequency can be from 2.5 to 5 MHz, and the crystal should be cut to a frequency in that range. In operation at the lower power levels, there is very little practical difference between the Colpitts and Pierce oscillators. As with all RF circuits, placement of components, layout, and circuit construction can be critical to successful circuit operation; also, "debugging" the circuit can be a time-consuming process. The circuit can be used as a Morse code transmitter for the amateur radio service by interrupting the power supply with a telegraph key. However, it must not be connected to a radiating antenna unless the operator holds the appropriate class of amateur radio license and operates it according to the rules of that radio service.

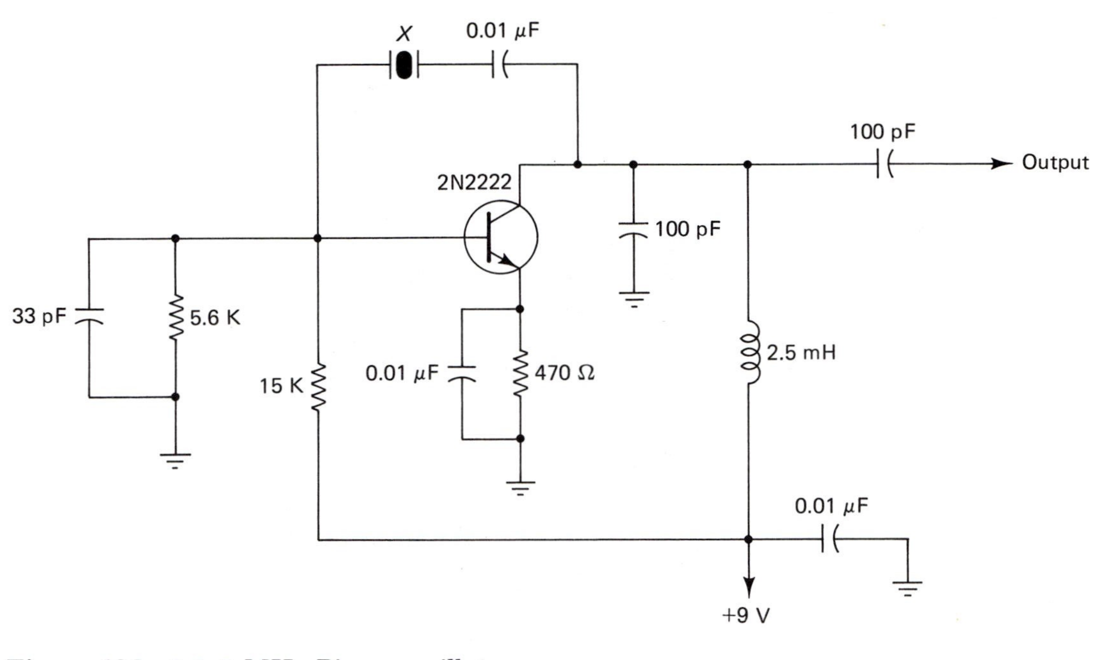

Figure 136 2.5–5 MHz Pierce oscillator

Oscillator, relaxation

A relaxation oscillator is based upon a semiconductor device rapidly charging and discharging a capacitor through a resistor. Most relaxation oscillators use a device known as a *unijunction transistor.* This device can be thought of as a double-base diode, in that it has two bases and an emitter, but no collector. Figure 137 shows a 2N2422 unijunction transistor used as a relaxation oscillator to produce a 1,500 Hz output. When power is first applied to this circuit, the 0.0022 μF capacitor charges through the 270 K resistor until the emitter voltage overcomes the reverse bias applied to the first base of the 2N2422. When current flows through the emitter, the resistance of the first base falls rapidly and the capacitor can discharge. This reduced resistance establishes a new bias point for the emitter junction. Then, more emitter current flows, a space charge builds up in the region of the first base, eventually shutting off the emitter current and allowing the capacitor to be recharged. The cycle then repeats, setting up the oscillation.

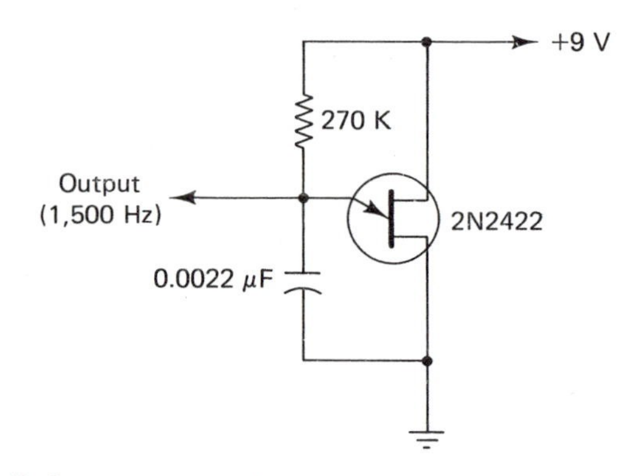

Figure 137 Relaxation oscillator

Oscillator, sine wave

The circuit in Figure 138 uses the common 741 op amp in a "twin-T" configuration to produce a 1,000 Hz sine wave output. The 100 K potentiometer is adjusted until the circuit "breaks into" oscillation. The output frequency of this circuit depends upon the values

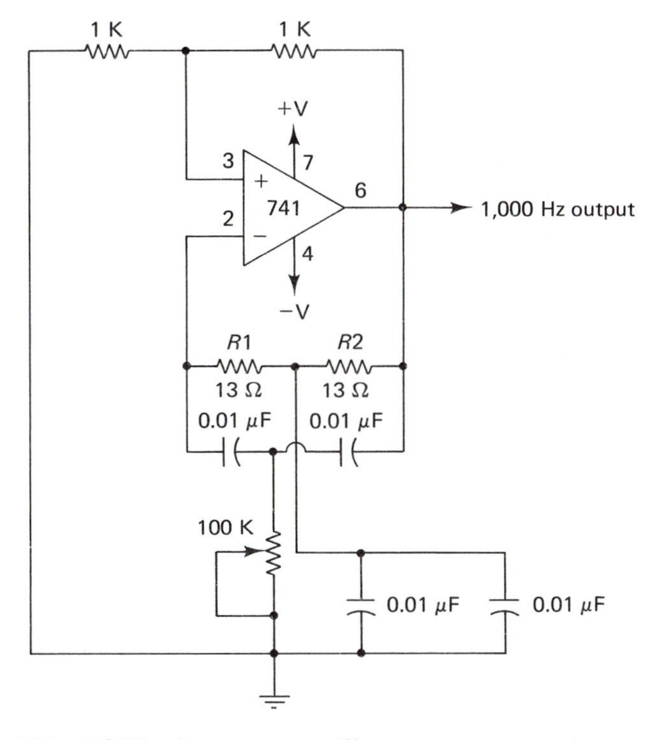

Figure 138 1 kHz sine wave oscillator

of resistors $R1$ and $R2$, which must be identical and inversely proportional to the oscillation frequency. The values used in the circuit shown, 13 Ω, produce the 1 kHz output. In general, these values should be in the range of 4.7 to 18 Ω. The exact frequencies obtained for given values of $R1$ and $R2$ will depend on a number of factors, including the precision of the various resistors and capacitors used.

Oscillator, square wave

A wide-range square wave oscillator can be built using the 741, as shown in Figure 139. This circuit can produce a stream of square wave pulses at frequencies from 100 to over 10,000 Hz. Pulse duration is controlled by adjustment of the 100 K and 1 M potentiometers, while linearity of the output can be controlled by the 10 K potentiometer. Fixed-value resistors may be substituted for the 100 K and 1 M potentiometers if only a single output frequency is

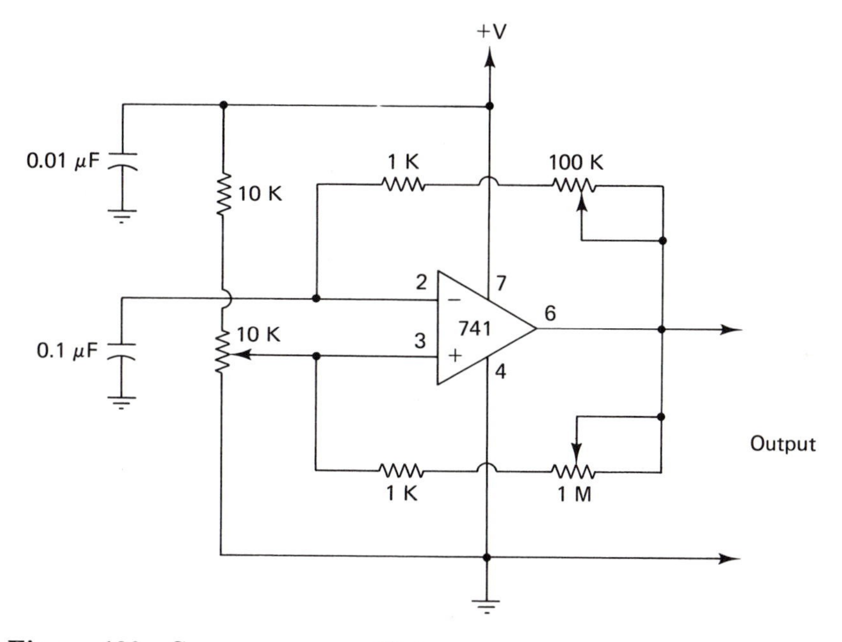

Figure 139 Square wave oscillator

needed. Unlike many applications of the 741, a ground connection point is used and output is taken between the 741's output (pin 6) and ground. The circuit can be "touchy" to construct for proper operation; in particular, the lead lengths between the two capacitors and the 741's pins must be kept short. If a single output frequency is all that is necessary, an astable multivibrator using a 555 IC or similar item is a better approach.

Oscillator, square wave and sawtooth

A versatile IC is the XR-2206 function generator device, which is capable of producing sine, square, triangle, sawtooth, and pulse waveforms of high quality. Figure 140 shows this circuit used in a square wave and sawtooth oscillator. The output frequency depends upon the values of resistors $R1$ and $R2$, as well as capacitor C. The formula is given by

$$\text{Output frequency} = \frac{2}{C} \times \frac{1}{R1 + R2}$$

The duty cycle of the circuit can be adjusted from 1 to 99 percent and is found by the formula

$$\text{Duty cycle} = \frac{R1}{R1 + R2}$$

The values of $R1$ and $R2$ must be in the range of 1 K to 2 M.

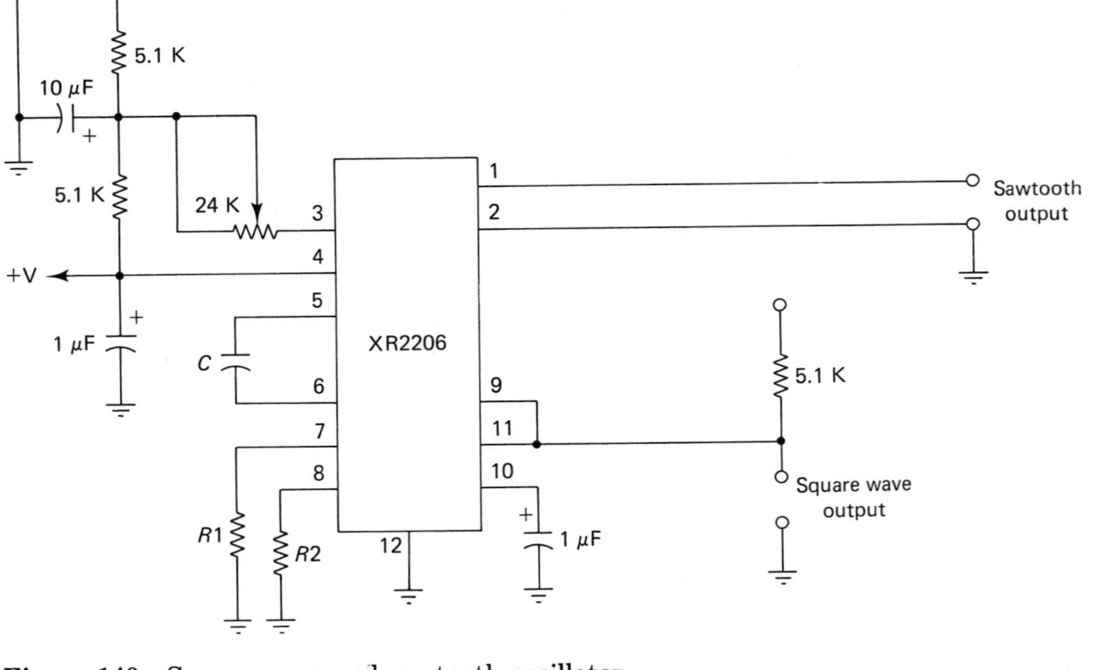

Figure 140 Square wave and sawtooth oscillator

Oscillator, triangular and square wave

Figure 141 shows a circuit that uses both halves of an SE/NE5532 dual-operational amplifier to generate triangular and square waves with high precision. The operating frequency of this circuit depends upon the values of resistors $R1$, $R2$, and $R3$, as well as capacitor C. The formula is

$$\text{Output frequency} = \frac{1}{4\,R1C} \times \frac{R3}{R2}$$

The dual-polarity supply voltage required may be as high as 22 V. The square wave output will maintain 50 percent duty cycle even if the amplitude of the oscillation is not symmetrical.

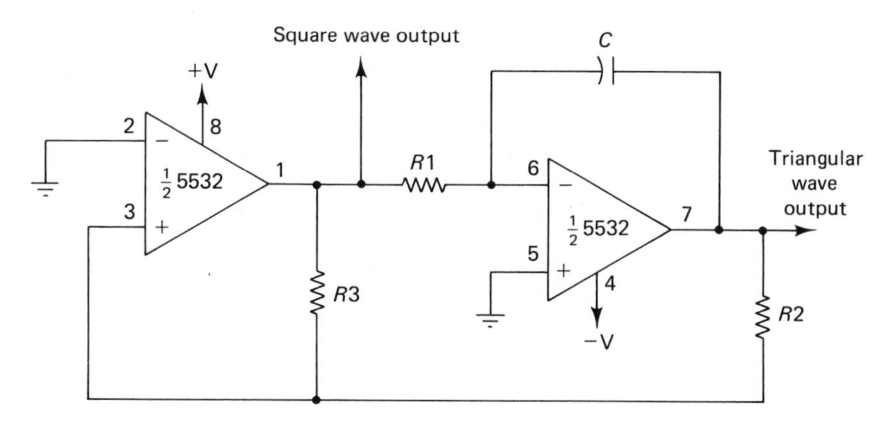

Figure 141 Triangular and square wave oscillator

Oscillator, ultrasonic

The output of the circuit in Figure 142 will be of interest primarily to dogs, since it is a square wave which is continuously swept from 30 to 45 kHz. There are also claims that the output can be effective in repelling certain types of insects. The heart of this circuit is a 555 timer configured as an extremely fast astable multivibrator. The speaker used is a piezo "supertweeter" type favored by some audiophiles or found in ultrasonic alarm systems.

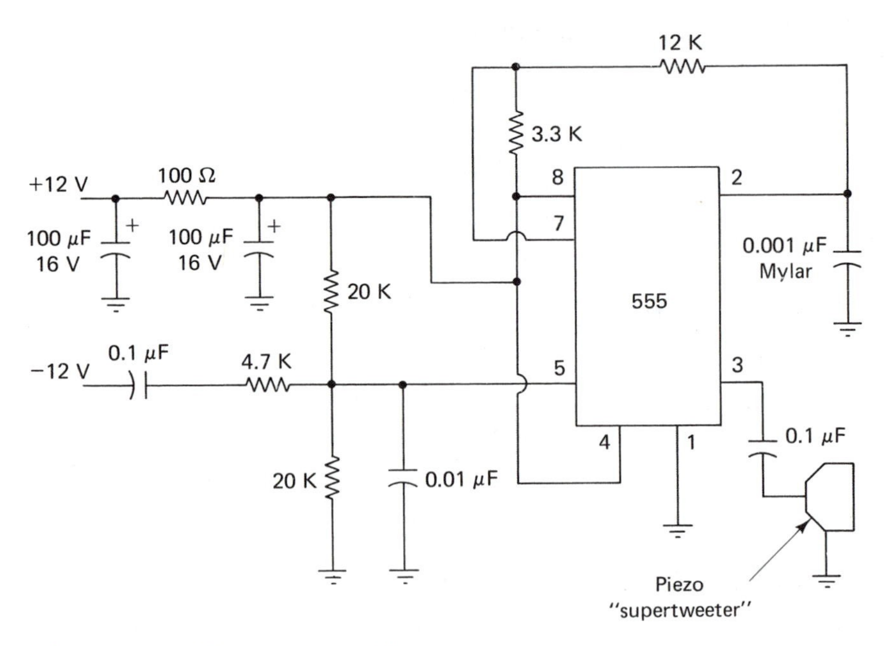

Figure 142 Ultrasonic oscillator

Oscillator, voltage-controlled

A voltage-controlled oscillator is one in which the output frequency varies according to an input voltage. The circuit in Figure 143 can accept an input voltage in the range of 0 to +15 V DC to produce a sine wave output from 0 to approximately 300 kHz. The two 741 op amps serve as buffers and "shapers" for the input and output signals, while the ICL8038 precision waveform generator produces the actual sine wave signal. The output of the first 741 IC is applied to pin 8 of the ICL8038, which is the FM sweep input. This causes the output frequency of the device to vary in accordance with the signal from the first 741 device. The 500 Ω potentiometer controls the high-frequency symmetry of the ICL8038, while the 100 K potentiometer across ground and the −15 V supply controls the low frequency symmetry. Adjustment of the 100 K potentiometer connected to pin 12 of the ICL8038 produces minimum distortion in the sine wave output. Note that a triangular waveform output is available at pin 3 and a square wave output at pin 9 of the ICL8038; however, the change in these with changes in input voltage will not be linear over their entire range.

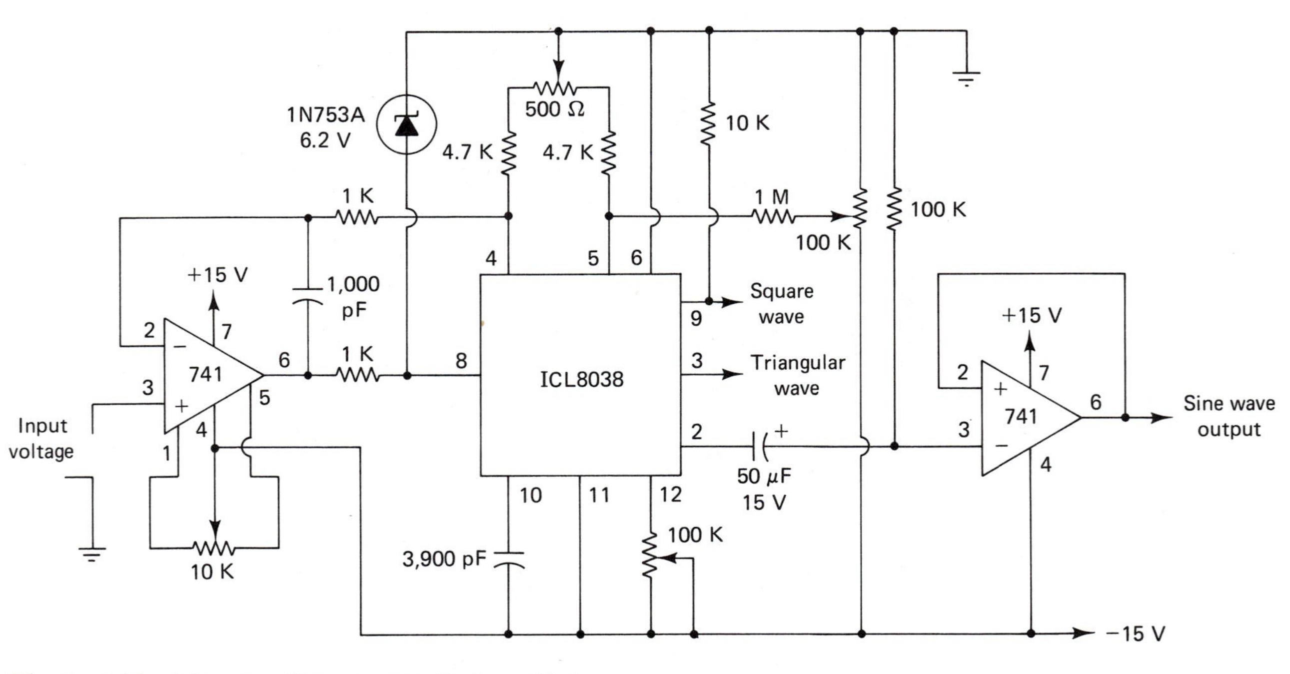

Figure 143 Linear voltage-controlled oscillator

Oscillator, Wien bridge

An Wien bridge oscillator is one in which a feedback loop with a 180° phase shift at the desired output frequency is used with barely enough gain in the loop to enable a self-sustaining oscillation to take place. This arrangement allows the Wien bridge oscillator to be an excellent source of low-distortion sine wave signals at low to middle frequencies. Figure 144 shows such a circuit implemented using one-

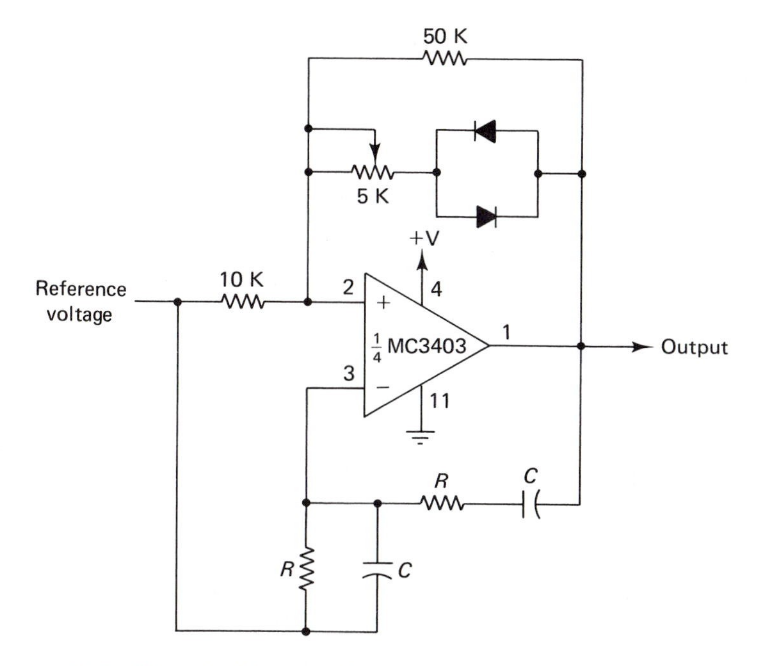

Figure 144 Wien bridge oscillator

quarter of an MC3403 quad low-power operational amplifier. The output frequency of this circuit is given by the formula

$$\text{Output frequency} = \frac{1}{2\pi RC}$$

in which the reference voltage is equal to one-half the supply voltage, which can range up to +36V. The 5 K potentiometer is used to control the amount of oscillation in the circuit, and should be set as low as possible to sustain the oscillation. The diodes can be 1N914s or other general-purpose types. A 1 kHz output version of this circuit would use 16 K resistors for R and 0.01 μF capacitors for C.

Oscillator, 85 kHz crystal

For highest stability, a quartz crystal should be used to control the frequency of an oscillator. Figure 145 shows an oscillator based on an LM311 operational amplifier. This circuit produces a stream of pulses at an 85 kHz rate controlled by a fundamental-mode crystal cut for that frequency. The circuit is also capable of operating over a wider frequency range: depending upon the components selected and their placement, the circuit may be able to oscillate at up to 100 kHz or down to a few kHz. The +5 V power supply makes the output of the circuit TTL compatible.

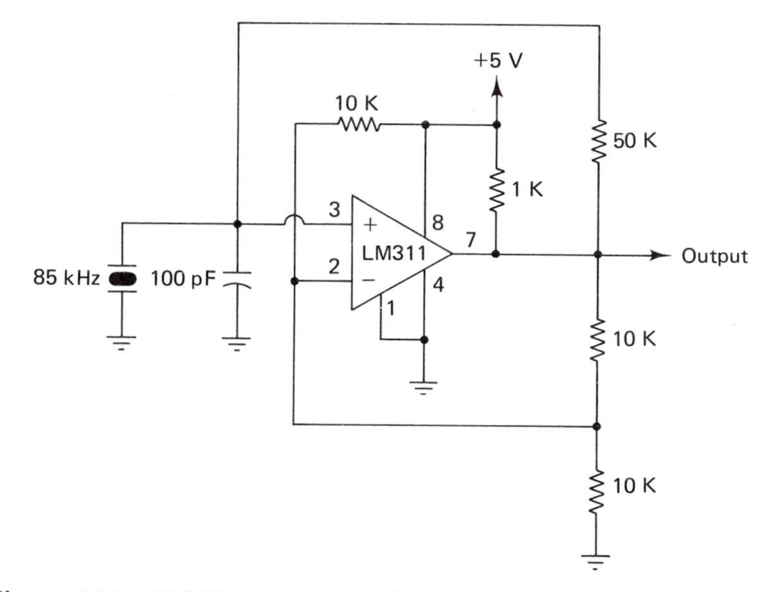

Figure 145 85 kHz crystal oscillator

Oscillator, 1.68 GHz

Transistors such as the 2N5108 now make it possible to construct oscillators for frequencies that once were restricted to laboratory experiments. Figure 146 shows such an application of the 2N5108, which produces 300 mW output at 1.68 GHz. *L*1 and *L*2 are $\frac{1}{16}''$ wide and 1.31″ long sections of microstrip board, while *L*3 is a $\frac{1}{16}''$ wide and 0.65″ long section of microstrip board. Both 25 pF capac-

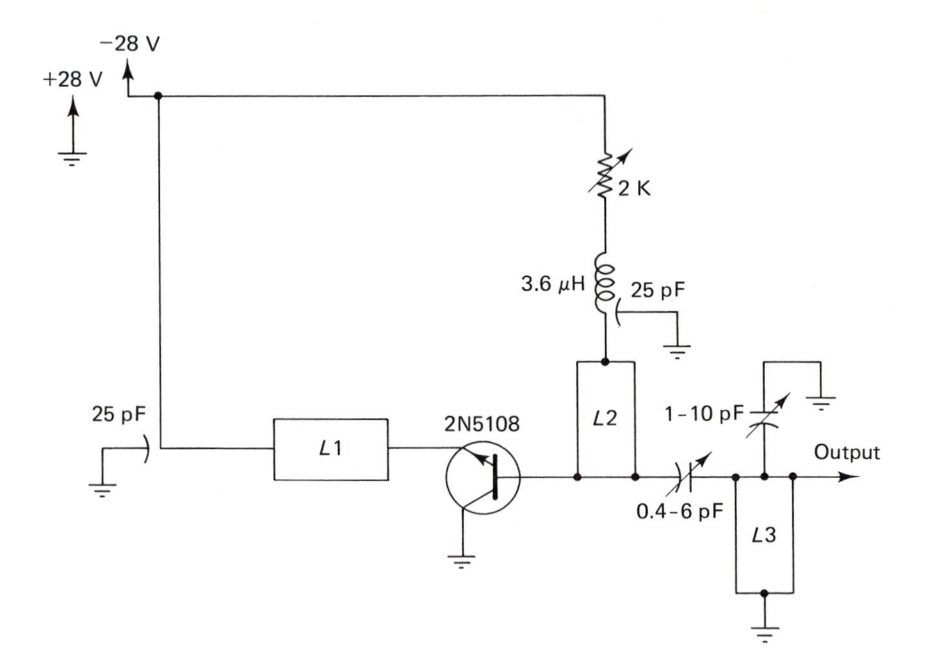

Figure 146 1.68 GHz RF oscillator

itors are "feedthrough" types. Construction of circuits operating at 1.68 GHz is usually critical, as even short lengths of part leads can develop unwanted inductances. A good RF ground is essential, as is proper placement of components. This circuit must not be connected to a radiating load without the appropriate operator and station licenses, and care must be observed in use since RF energy at 1.68 GHz can cause damage to eyes and other sensitive human tissues. The emitter-to-base voltage should be approximately 2 V when the circuit is operating properly.

p

Parity checker

Parity checking is a method of testing the accuracy of data words by adding an extra bit, known as a parity bit or check bit, to a data word. The parity bit is compared to the sum of all high-level (1) bits found in the data word. The checking function can be performed by IC devices developed specially for the purpose, such as the 74LS280. In Figure 147, the bit at input A will be the parity bit, and the

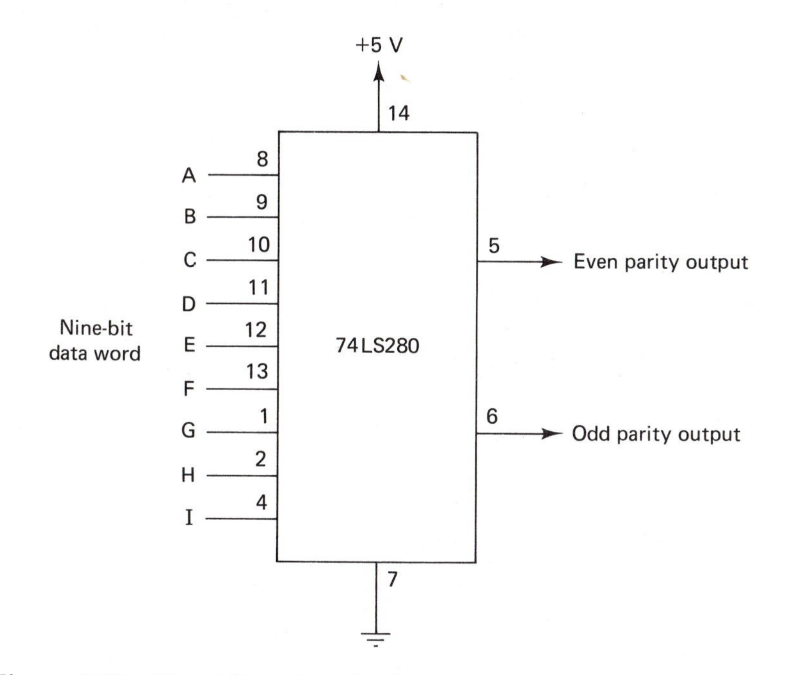

Figure 147 Nine-bit parity checker

remaining bits make up an eight-bit data word. The even parity output will be high and the odd parity output low when the sum of the high-level inputs is an even number. By contrast, the even parity output will be low and the odd parity output high when the sum of the high-level inputs is an odd number.

It is possible to cascade more than one 74LS280 to perform a parity check on larger data words. The circuit is Figure 148 will compare a 17-bit data word consisting of 16 bits of data and a single parity bit. In cascading, the odd parity output is connected to the A input (pin 8) of the next 74LS280.

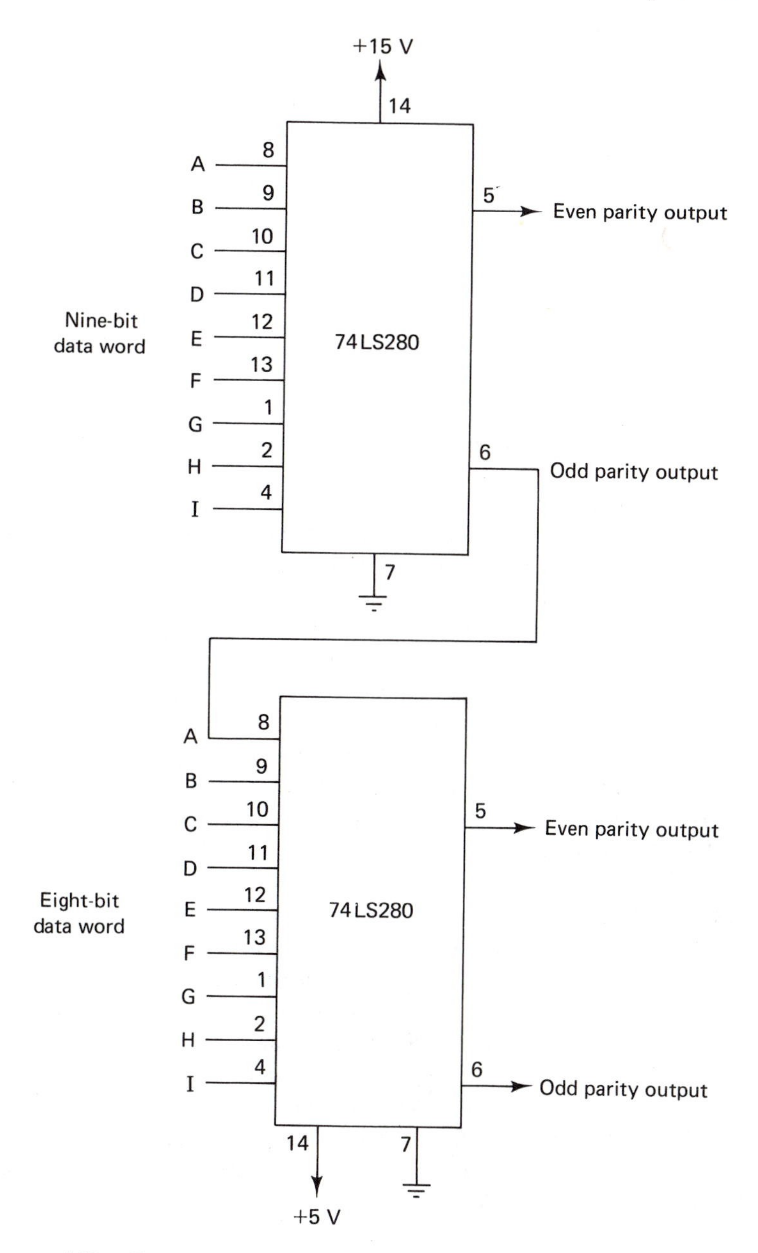

Figure 148 Seventeen-bit parity checker

Phase-locked loop

A phase-locked loop (PLL) is a circuit in which the output of a voltage-controlled oscillator (VCO) is compared to the output of a reference oscillator in a phase detector. If the phase of the VCO's output differs from the phase of the reference oscillator's output, a correcting voltage is produced by the phase detector and fed back to the VCO so that its output goes into phase with the reference oscillator's output. Thus, a PLL can track an input signal applied to it. The range of frequencies centered around a specific frequency over which the PLL can track frequency changes is known as the *tracking range* of the PLL. PLL IC devices are available and are widely used for such purposes as FM and FSK demodulation, as well as frequency synthesis.

For some purposes, a PLL constructed from other components may be needed. The circuit in Figure 149 is a PLL built out of an XR-2228 multiplier/detector device and an XR-2209 precision VCO IC. This circuit has much better temperature stability than most PLL ICs and has a tracking range of 12:1 centered on 840 kHz. In operation, the 10 K potentiometer between the XR-2228 and the XR-2209 is adjusted until the circuit locks on the input signal and follows its variations. The VCO output (at pin 8 of the XR-2209) is a triangular waveform, in contrast to the demodulated output taken from the XR-2209 itself. The output of the circuit may be swept from 120 kHz to 1.4 MHz.

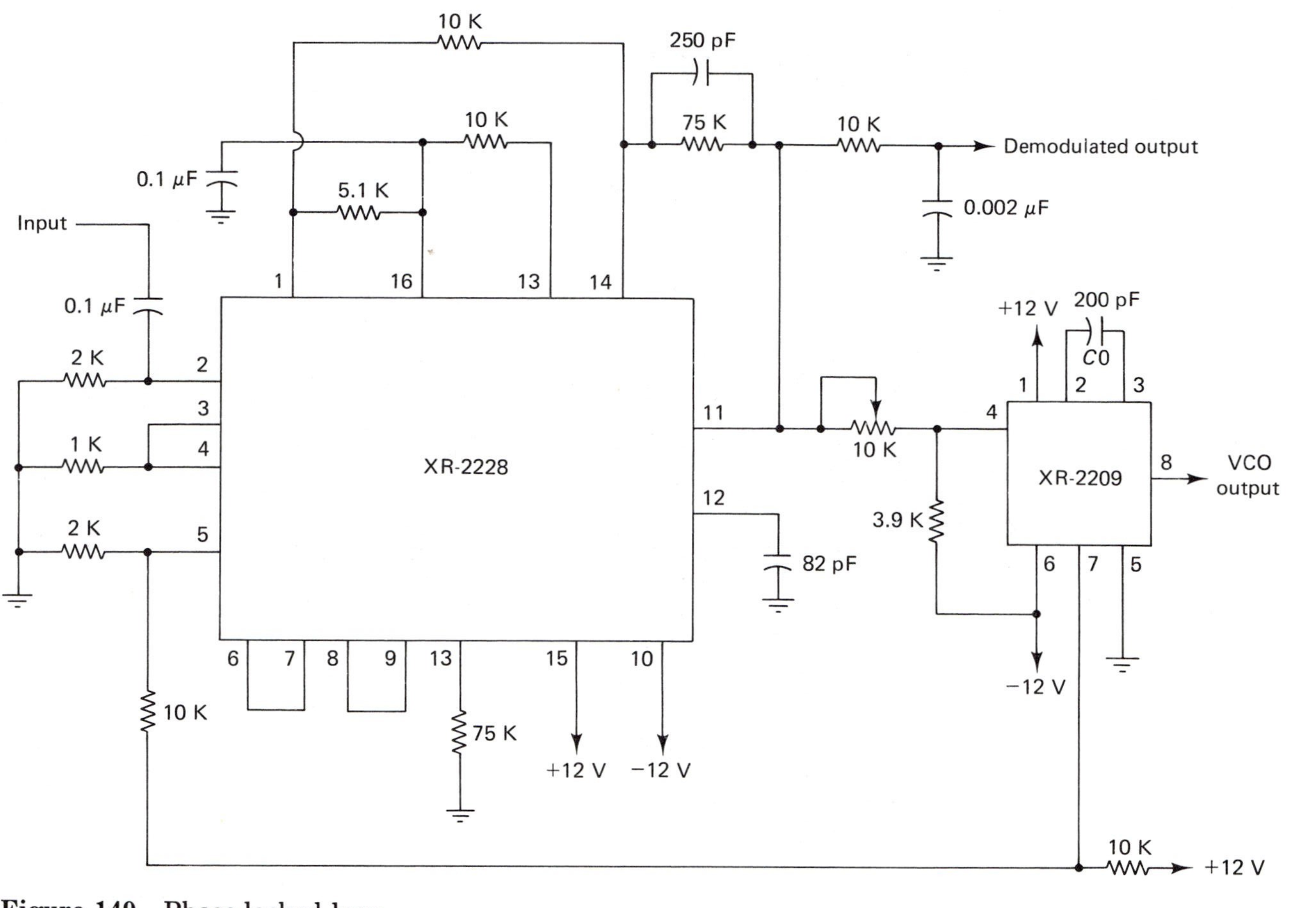

Figure 149 Phase-locked loop

Phase shifter, digital

The circuit in Figure 150 will take a clock input signal, divide it in half, and produce two clock signals of equal frequency 180° out of phase with each other. For example, feeding a 100 kHz clock signal to the circuit will produce two 50 kHz outputs with opposite logic levels, i.e., when the output of one is high the other will be low, and vice versa. The heart of this circuit is a 4013 CMOS dual D flip-flop device, and the clock signal input must be a CMOS-compatible clock. Naturally, the outputs will be CMOS-compatible as well.

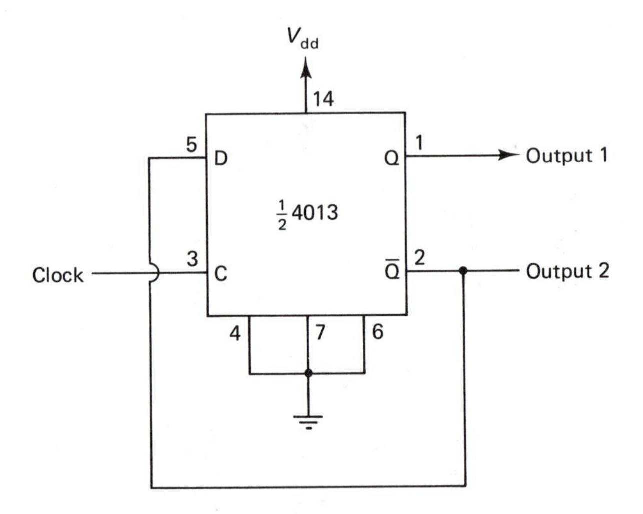

Figure 150 Digital phase shifter

Power supplies, AC

These circuits take an AC power line voltage in the range of 110 to 120 V (noted as 120 V AC in the schematic diagrams) and produce a steady DC voltage from it. They all use *step-down* transformers, which take the AC line voltage and produce an output voltage that is much less, usually only a few volts. This reduced voltage is still AC, however, and must be rectified to produce a DC voltage. Each circuit uses *bridge* rectification for this purpose. A bridge rectifier consists of a network of four diodes or a single integrated module containing four diodes in the proper configuration. Regardless of which type it is, the bridge rectifier allows both the positive and negative halves of an AC waveform to be used; the negative half of the cycle is "flipped over" to become a positive voltage. The resulting output of the bridge rectifier is a *pulsating* DC voltage which varies in amplitude. As a result, capacitors and, in some designs, inductors are used to smooth the output into a relatively constant-level DC voltage.

As predicted by Ohm's law, the output voltage produced by a power supply decreases as the current demanded by a load increases. To overcome this problem, *voltage regulation* circuits are often used. A voltage regulator takes an input voltage in excess of the desired output voltage and produces an output voltage that remains constant regardless of the changes in output current demand. For example, suppose a steady +5 V output voltage is needed. Then a voltage regulator circuit would take a higher voltage, such as +8 V, and use it to provide a steady +5 V output over a broad range of current demands. Voltage regulators ICs, such as the 78xx-series, are now commonly used for such functions.

Figure 151 shows a power supply that is highly suitable for powering TTL circuits. The output is a constant +5 V at up to

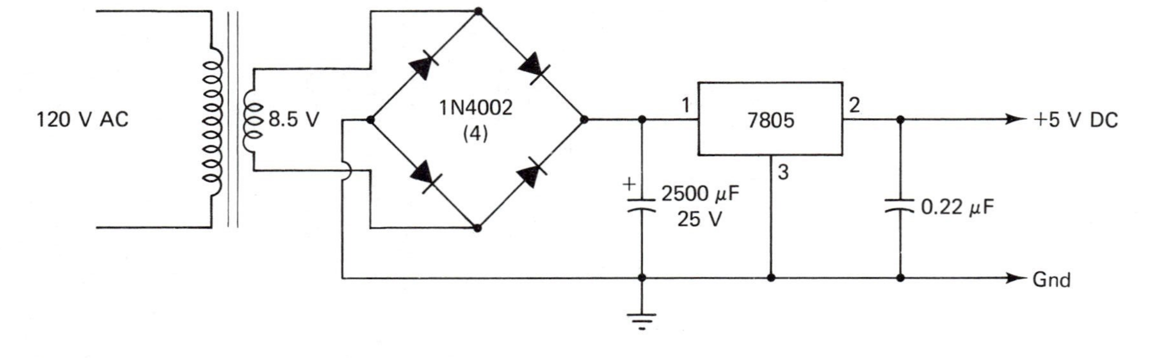

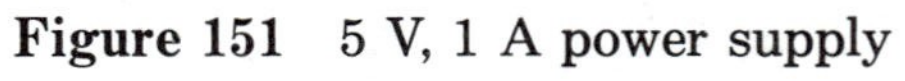

Figure 151 5 V, 1 A power supply

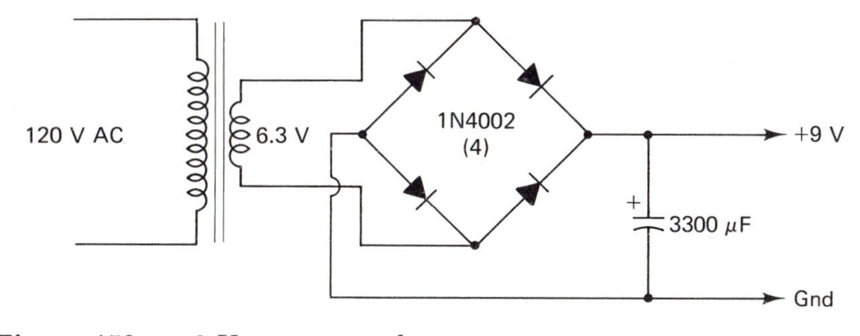

Figure 152 +9 V power supply

1 A thanks to the 7805 voltage regulator. The power transformer should have a secondary rated at 8.5 V.

A simpler circuit that can be used with linear devices is shown in Figure 152. This circuit delivers +9 V, suitable for single-supply operation of op amps and similar ICs. Note, however, that it is an unregulated supply and thus is not suitable for applications in which current demand will be heavy or variable. It should be fine for simple circuits involving only a single device. The transformer secondary is rated at 6.3 V.

A versatile supply suitable for operation of CMOS and single-supply linear devices is shown in Figure 153. This circuit can deliver a DC voltage from +8 to 14 V by adjustment of the 1 K potentiometer. The 7805 voltage regulator IC again provides excellent regulation of the voltage selected. The transformer secondary is rated at 12 V.

A greater voltage output range is provided by the circuit in Figure 154. This circuit uses the 317 voltage regulator IC to deliver an output from +1.5 to 37 V at up to 2 A of current. The circuit is suitable for powering TTL, CMOS, or linear devices, and ample output current allows it to power complex circuits involving several devices. The exact output voltage is set by the 5 K potentiometer. The bridge rectifier is composed of four diodes rated at 4 A and 50 V peak inverse. The transformer secondary must be rated for 25 V at 2 A.

Sometimes a dual-polarity voltage is needed. One approach to obtaining this is shown in Figure 155, which produces a dual-polarity 6 V output and provides a ground point as well. The power transformer has a secondary rated at 12.6 V and is center tapped.

A dual-polarity supply useful with op amps is shown in Figure 156. The output of this circuit is 12 V and is fully regulated: the

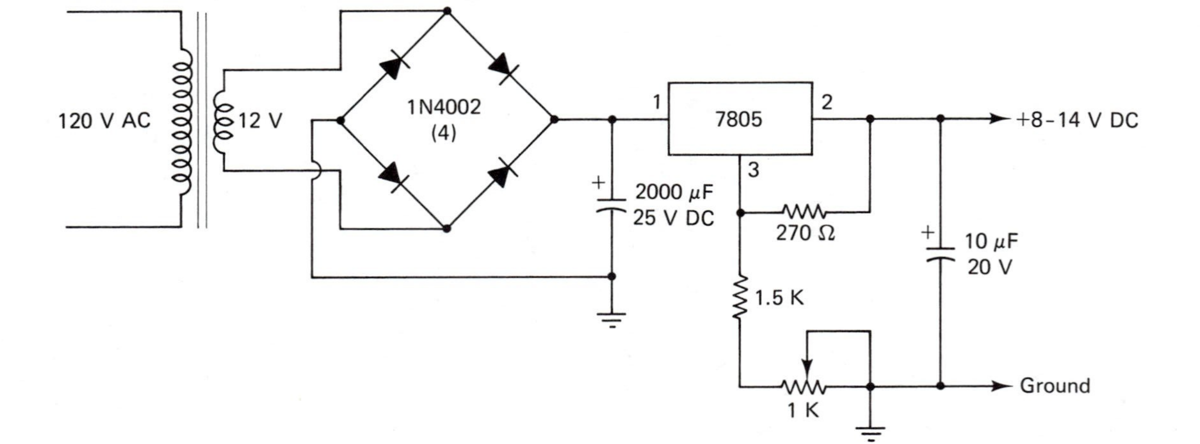

Figure 153 +8–14 V DC power supply

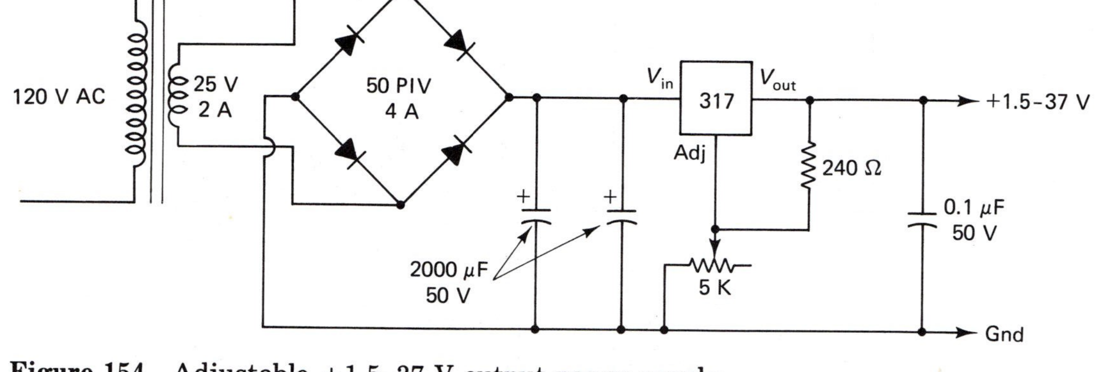

Figure 154 Adjustable +1.5–37 V output power supply

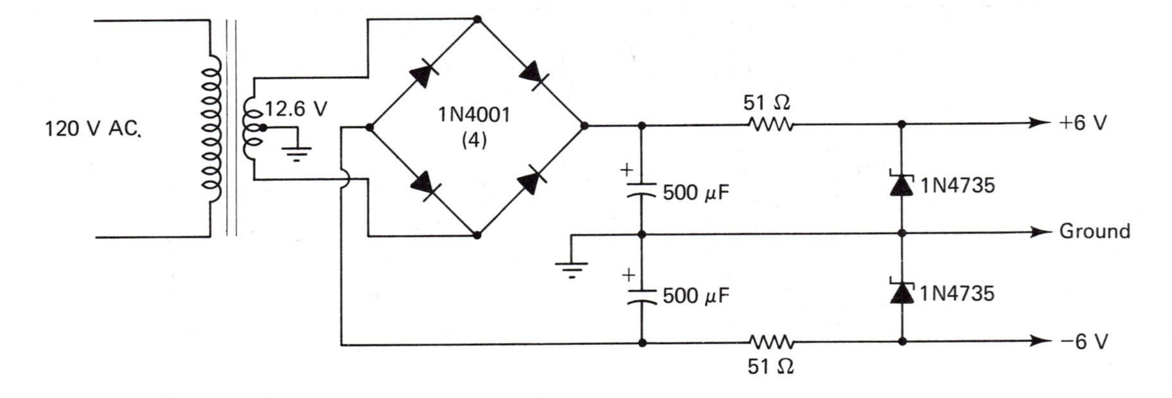

Figure 155 Dual-polarity 6 V supply

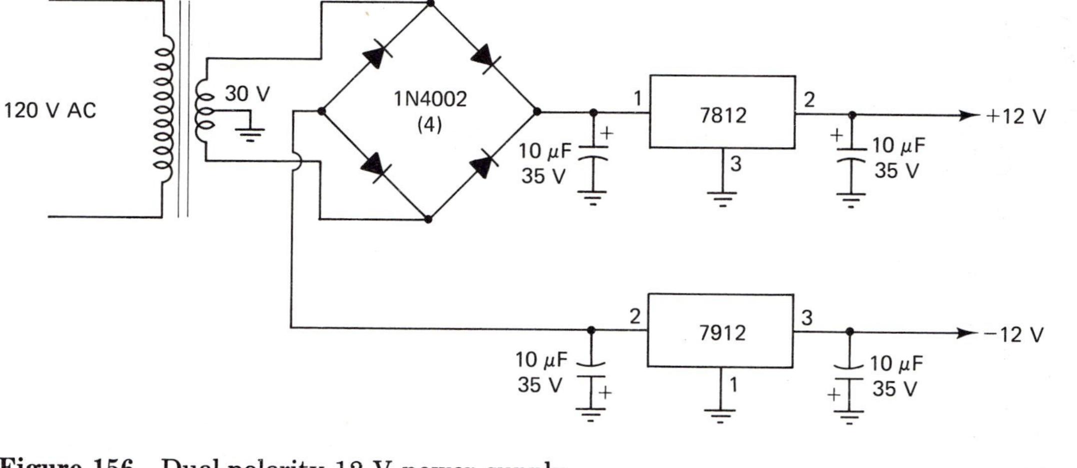

Figure 156 Dual-polarity 12 V power supply

7812 regulates the positive-polarity output, while the 7912 regulates the negative-polarity output. The power transformer has a secondary rated at 30 V and is center tapped to ground. Care must be taken with respect to the polarity of the electrolytic capacitors used.

Since all of the AC power supplies discussed involve a connection to a standard AC power line, care must be taken in the construction and checking of these circuits for maximum safety.

Power supplies, battery

For linear devices that require a single-polarity voltage or CMOS circuits, a standard 9 V battery will do as a power source. However, it and other batteries are not suitable per se for powering TTL devices or circuits that need dual-polarity voltages.

Figure 157 shows a circuit that produces +5 V for powering TTL. The 6 V battery source can be provided by four 1.5 V batteries. The 1N4001 diode creates a constant voltage drop sufficient to provide +5 V at its output. Since TTL is a "power hungry" logic family, the output of this circuit will soon drop below 4.75 V (the minimum TTL supply voltage) in many cases.

Figure 158 shows how to configure two 9 V batteries to produce a dual-polarity 9 V supply. The two batteries share a common ground connection point.

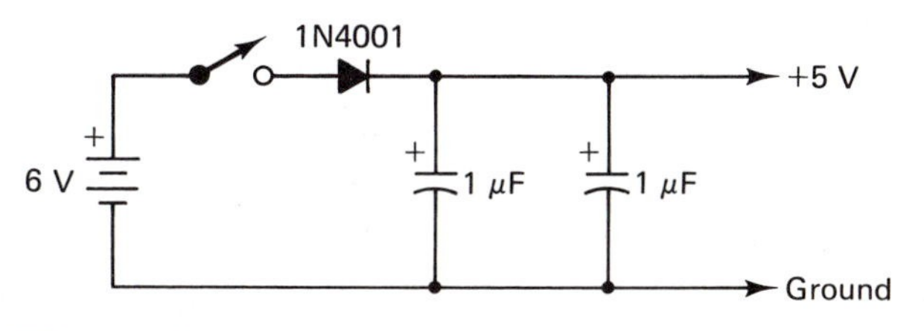

Figure 157 5 V DC supply

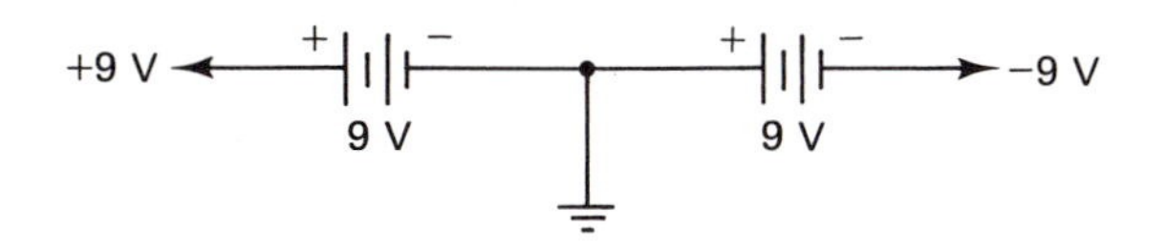

Figure 158 Dual-polarity 9 V supply

Preamplifier, microphone

The output from a microphone needs some amplification before it can be applied to an amplifier, mixer, or transmitter. Figure 159 shows a circuit designed for use with dynamic microphones of low to medium impedance; an 8 Ω speaker may also be used. The op amp used is one section of a TL084 quad JFET-input device. The gain of the circuit is set by adjustment of the 1 M potentiometer. If a fixed amount of gain is desired, the potentiometer may be replaced with a resistor and the final gain of the circuit will depend upon the value of the substitute resistor divided by 10 K.

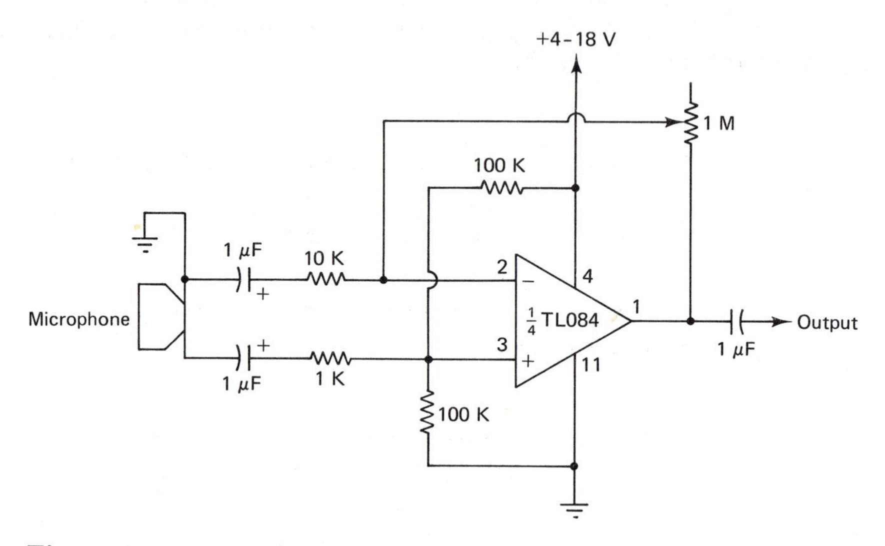

Figure 159 Microphone preamplifier

Pulse generators

A pulse generator can be used to provide clock signals for logic devices, drive LEDs or speakers, or provide a steady source of reference signals. Both of the pulse generators to be described provide TTL-level (+5 V) pulses.

The circuit in Figure 160 is implemented using both halves of an MC1458 dual-operational amplifier. The pulse frequency depends upon the value of capacitor *C* as shown in the following table:

Capacitance	Frequency
0.001 μF	5,872 Hz
0.010 μF	660 Hz
0.100 μF	51 Hz
1.000 μF	8 Hz

A versatile pulse generator that uses the 555 timer is shown in Figure 161. The output frequency of this circuit typically ranges from 100 Hz to over 10 MHz depending upon the setting of the 1 M potentiometer. If powered by a +5 V source, the output will be TTL-compatible.

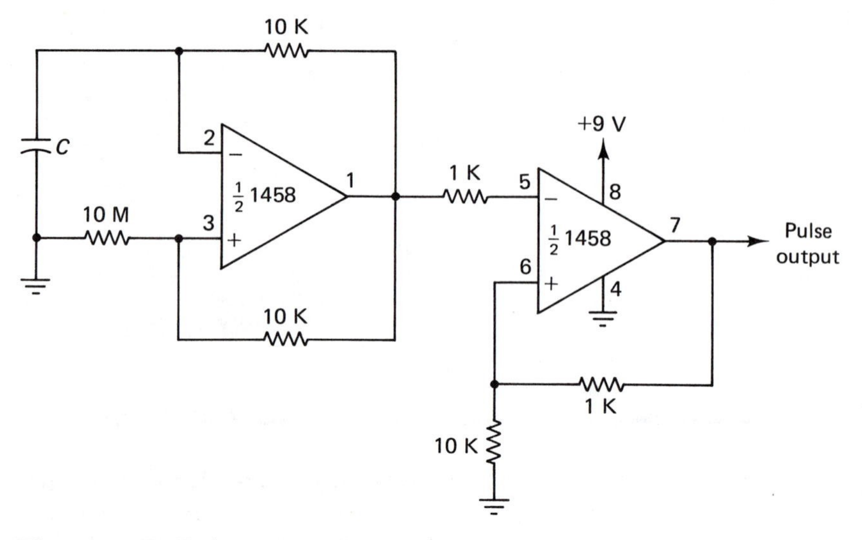

Figure 160 Pulse generator

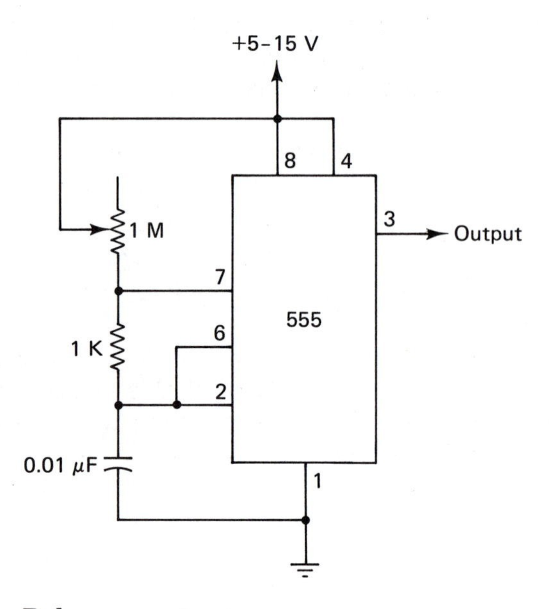

Figure 161 Pulse generator

Pulse shaper

Digital logic has a problem with clock or input signals which are "noisy." Noisy digital signals are those which are poorly formed, with irregular tops and sides, or those in which there are smaller amplitude pulses between the desired signals. Noisy clock and input signals can cause irregular operation of logic ICs, particularly when the noise is near a device's threshold for determining when a signal is at a low or high logic level. The solution to the problem is to apply noisy signals to an optimized comparator which replicates the input signal as a clearly formed output signal. Such a circuit is known as a signal shaper and is shown in Figure 162 as applied to processing

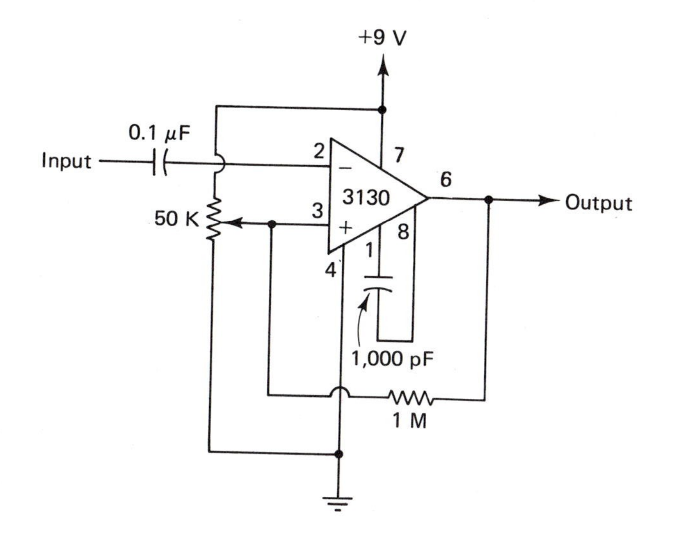

Figure 162 Pulse shaper

CMOS-level signals. The CA3130 is a BiMOS op amp, meaning that it uses a combination of bipolar and MOS transistors. The 50 K potentiometer is used to adjust the circuit for best operation at a given input frequency, and the circuit may be used for input frequencies up to 50 kHz. Other types of op amps may be used with this general configuration for TTL-compatible output and the like.

q

Q-multiplier

A Q-multiplier is essentially a special type of bandpass filter which operates in the intermediate frequency (IF) amplifier section of a radio receiver, usually at 455 kHz. A Q-multiplier increases the selectivity of the receiver by having a sharp peak response at one frequency in the receiver's bandpass while greatly attenuating other frequencies in the bandpass. In addition, the circuit can also reject

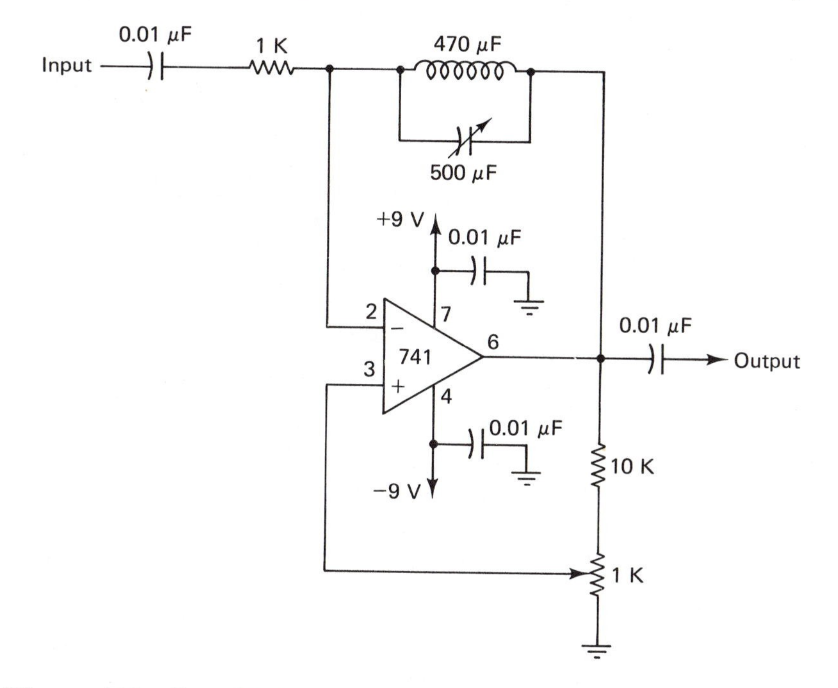

Figure 163 Q-multiplier

a specific frequency on which an interfering signal is present. Figure 163 shows a simple Q-multiplier built around a 741 op amp. Main tuning of the circuit is by the 1 K potentiometer, which adjusts the frequency that the circuit passes with the least attenuation. The 500 μF trimmer capacitor across the 470 μH inductor is used during circuit adjustment and testing; once the setting for proper operation is found, it is left alone.

r

Receiver, DTMF

"DTMF" is an acronym for *dual-tone multifrequency,* a tone dialing system used with pushbutton telephones. (The system is perhaps best known under the AT&T trademark of "Touch Tone.") Besides telephone dialing, DTMF tones can be used for a variety of signaling and control purposes, including transmission of digital data. Figure 164 shows a circuit which can produce a four-bit binary output from DTMF input tones. This circuit uses an S3525A DTMF filter IC and an MK5103 DTMF decoder device; the internal clock of the S3525A is controlled by a 3.58 MHz "color burst" crystal, and the S3525A clock serves as the reference for the MK5103. The 50 K potentiometer connected between pins 11 and 13 of the S3525A adjusts the bandpass of the DTMF filter.

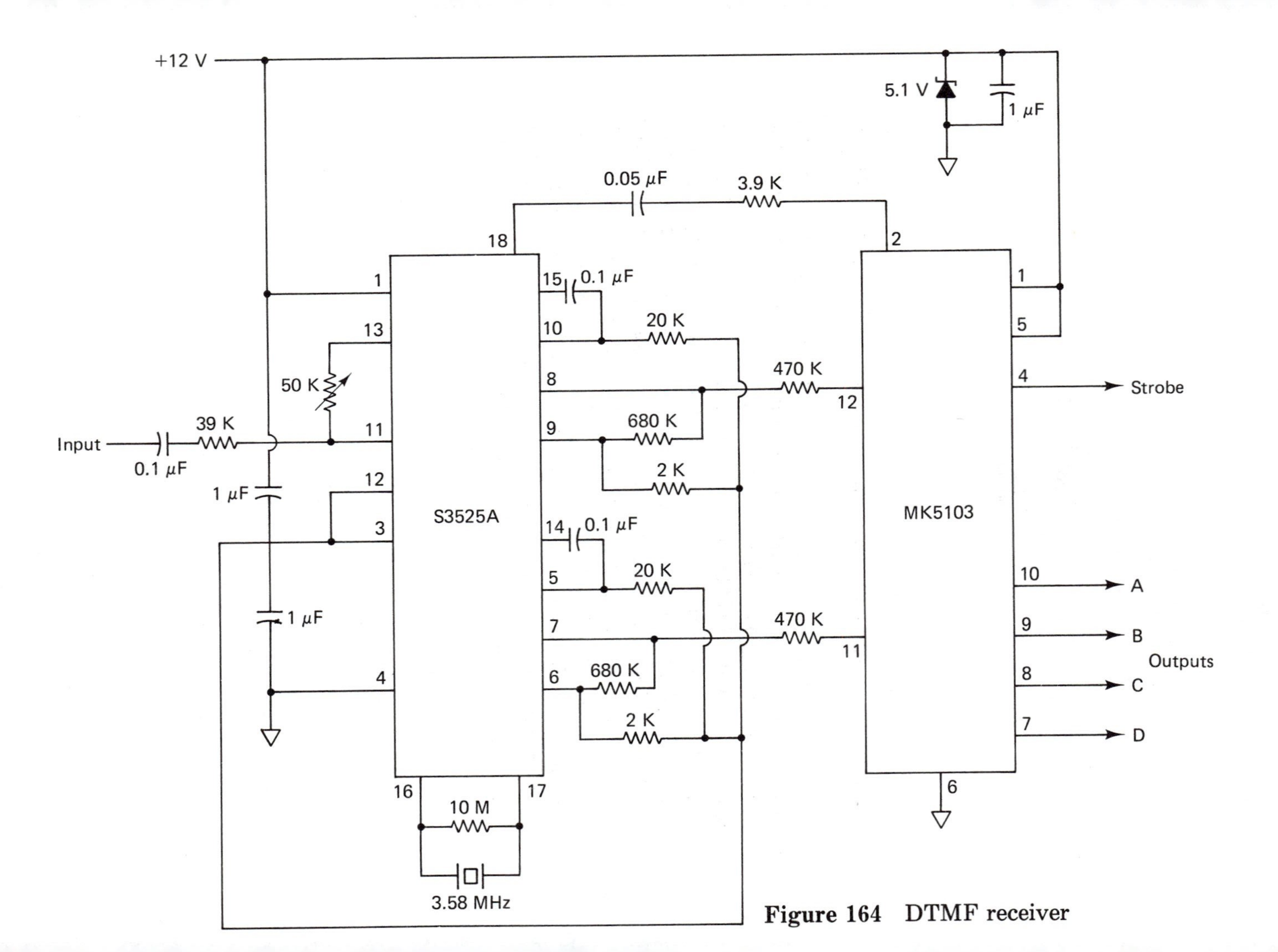

Figure 164 DTMF receiver

Receiver, infrared

The circuit in Figure 165 makes use of an MC3373 infrared remote control amplifier-detector IC and an SFH206 infrared receiver diode. The MC3373 combines a high-gain amplifier section along with pulse-shaping circuitry to provide a clean square wave output. The detector section is an envelope type optimized for PCM signals. The input signals to the circuit are generally in the form of bursts. The circuit was originally developed for use in TV remote control systems, but can be adapted for other purposes.

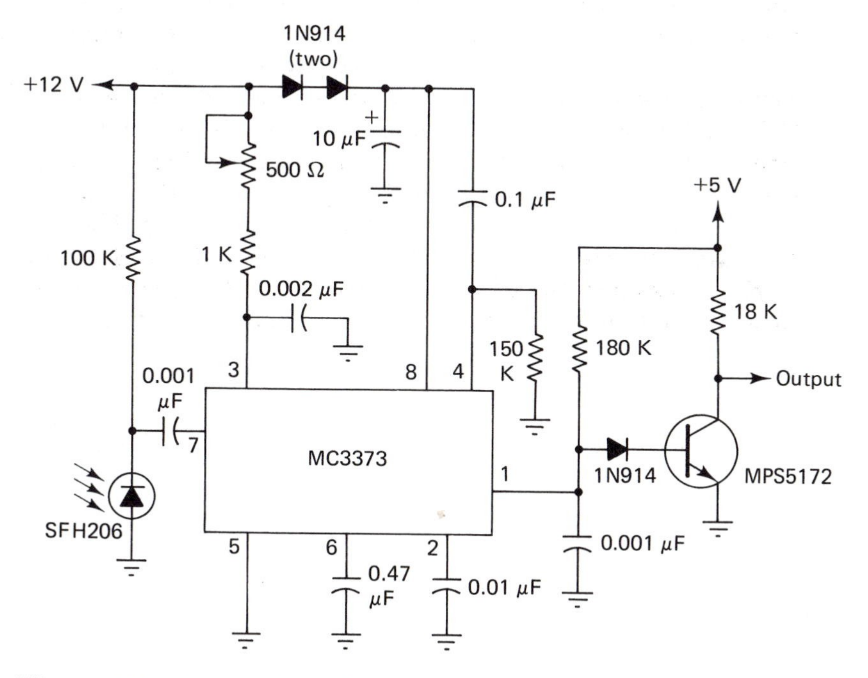

Figure 165 Infrared receiver

Receiver, ultrasonic

Ultrasonic audio frequencies are frequently used in remote control and signaling operations. Figure 166 shows an elaborate ultrasonic receiver with an input frequency centered on 40 kHz based on the MC14458 remote control receiver IC. The MC14458 delivers an eight-bit data word output, with the most significant bits at pins 8 through 11 and the least significant bits at pins 13 through 16. In addition, there are address outputs at pins 19 through 22. Since this circuit was developed for use in TV remote control systems, there are also outputs for volume, on/off, automatic fine tuning (AFT), and other TV functions. However, the circuit can also be used with communication receivers, video games, and similar devices. The MC14458 is preceded by an amplifier which uses three inverters from an MC14069 device; a fourth inverter provides an oscillator input to pin 1 of the MC14458. The frequency of this oscillator is set to 500 kHz by a ceramic resonator (labeled CR). The 100 pF capacitor connected to the two 1N914 diodes and 1 K resistor should be placed as close as possible to pin 2 of the MC14458. The ultrasonic "microphone" can be a piezo or ceramic ultrasonic speaker.

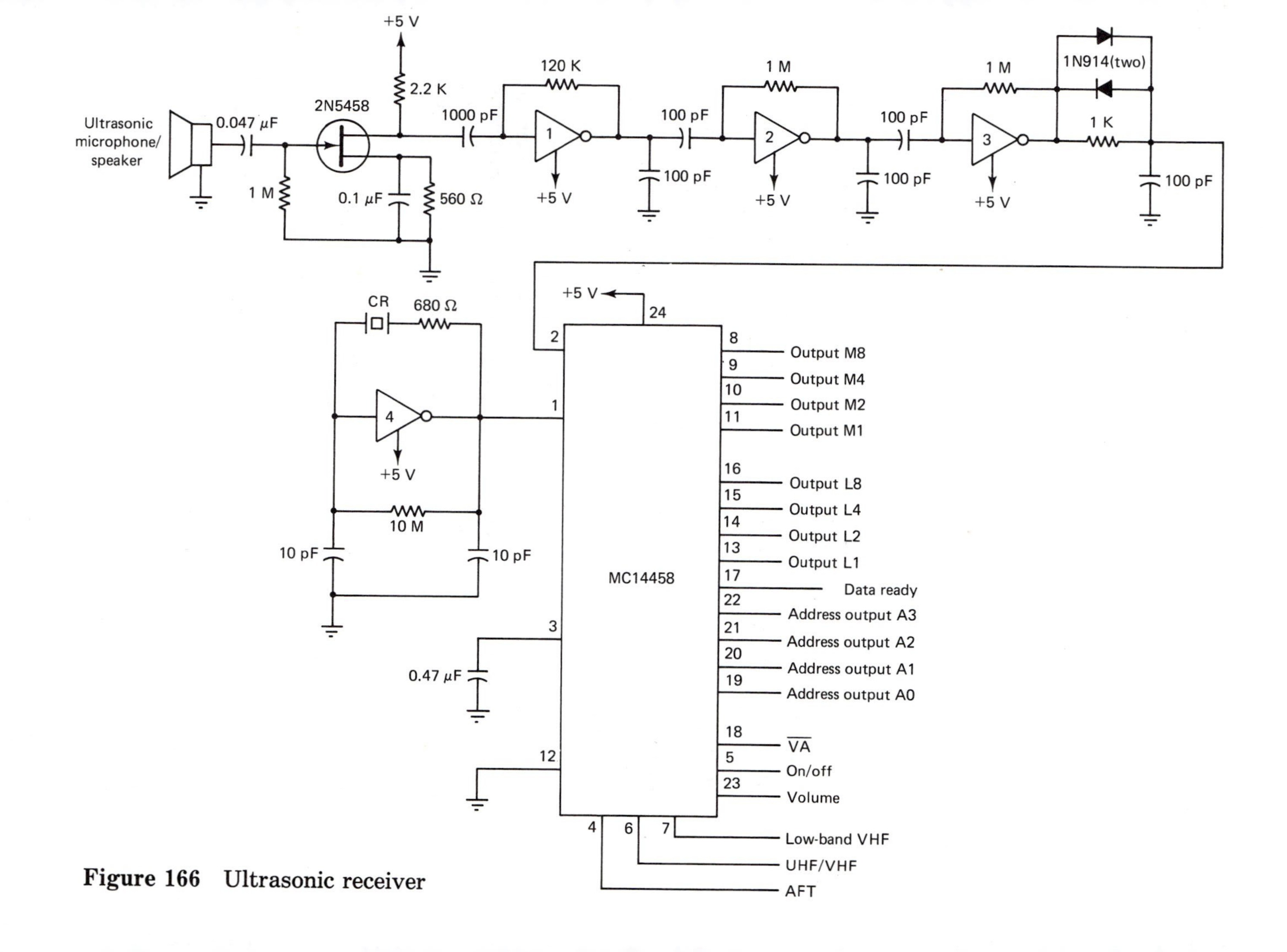

Figure 166 Ultrasonic receiver

Rectifiers

Customarily, rectifiers are simple diode networks. However, there may be times when "precision" rectification is needed, or an AC input may be of unusually high frequencies. In these situations, an active rectifier using an op amp is the preferred approach. Figures 167 and 168 show two rectifier circuits that use the SE/NE5535 dual high-slew-rate op amp. As with all op amp circuits, the input signal voltage must not exceed the supply voltage (18 V maximum in this case).

The circuit in Figure 167 will work with AC input signals of up to 10 kHz. It is a half-wave rectifier, meaning that only the positive half of each AC input cycle is used. For positive halves, the circuit gain is 0; for negative halves, it is −1. By reversing the two diodes, the output polarity of the circuit is reversed. The diodes can be any general-purpose types having voltage ratings equal to the input voltage.

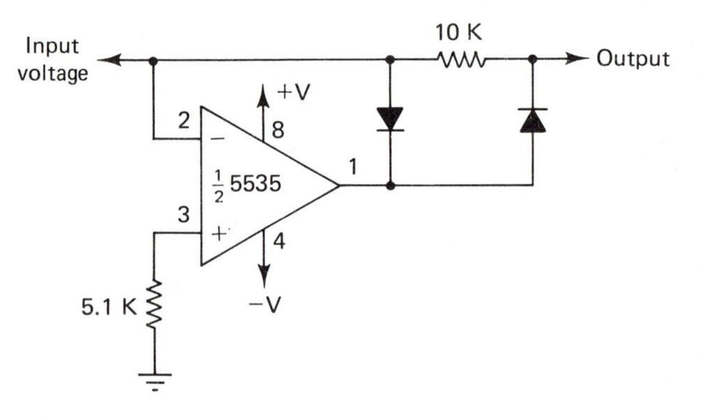

Figure 167 Half-wave rectifier

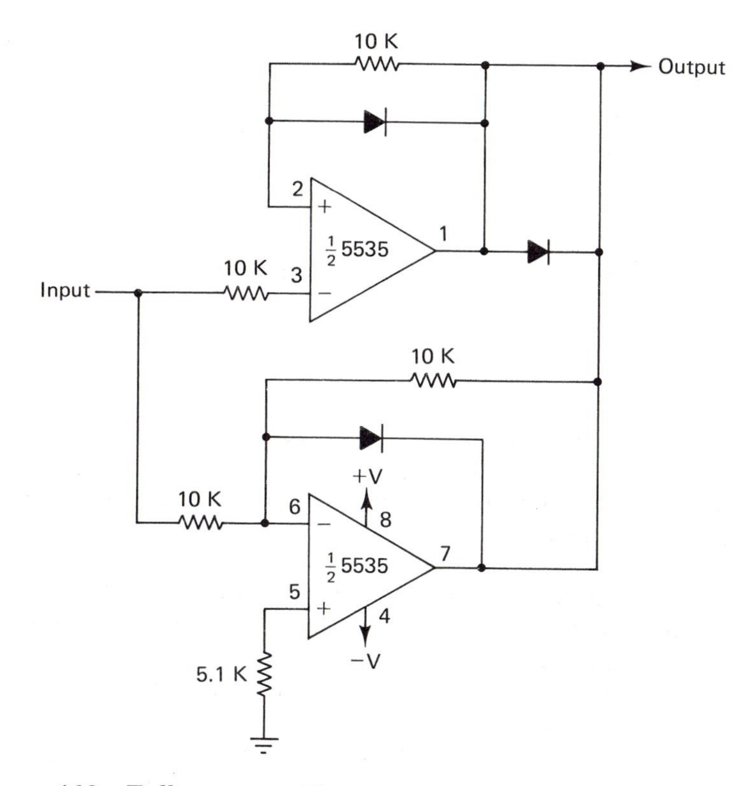

Figure 168 Full-wave rectifier

Both halves of the 5535 are used to make a full-wave rectifier, as shown in Figure 168. The performance is similar to that of the half-wave rectifier, and reversing the polarity of the diodes will reverse the polarity of the output voltage. Since the output will not sink heavy currents, the load applied should be referenced to ground or a negative voltage.

RF probe

Conventional voltmeters are designed to measure DC voltages and cannot be accurately used with AC signals. This problem is especially acute at radio frequencies. The solution is to use a circuit between the RF voltage source and voltmeter to produce an equivalent DC voltage to the RF voltage, as shown in Figure 169. The DC output of this circuit is equal to the RMS value of the RF voltage, and is accurate to within 10 percent from 50 kHz to 250 MHz. The impedance of the probe is approximately 6000 Ω. A good ground-point connection is essential to prevent stray RF from getting into the voltmeter and causing inaccurate readings or damage.

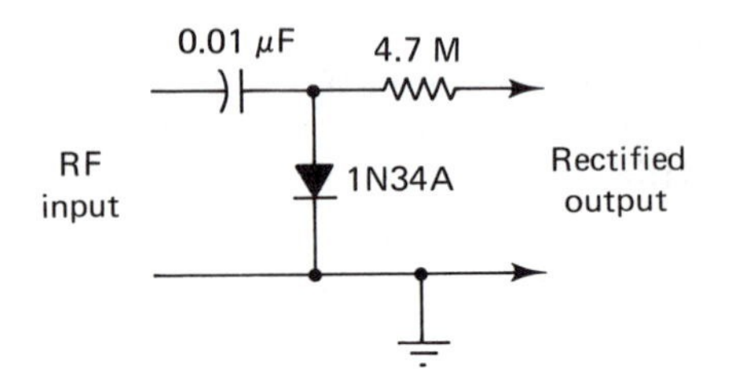

Figure 169 RF probe

S

Sample-and-hold

A sample-and-hold circuit is a circuit that samples an analog input and stores it, usually using a capacitor, for subsequent reading or conversion to digital form. The usual method of accomplishing this is to use an analog switch IC in conjunction with an op amp which offers gain to the charge stored in a capacitor.

Figure 170 shows a sample-and-hold circuit that uses a 4066 quad CMOS analog switch and half of a 353 dual JFET-input op amp. The operation of the circuit is controlled by a sample/hold input signal. When this input is high, the circuit will sample the input voltage signal. But when the sample/hold input goes low, the sample is held and the circuit will ignore any further input voltage signals until the sample/hold input goes low again. The output of the circuit is CMOS compatible.

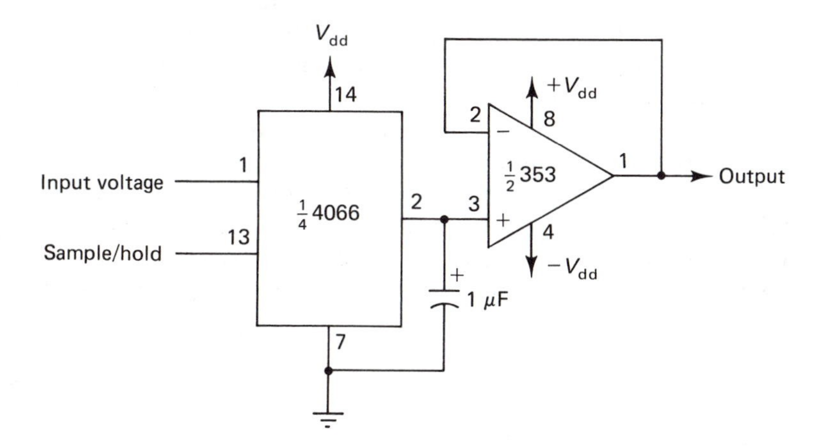

Figure 170 Sample-and-hold

SCA decoder

"SCA" stands for *subsidiary communications authorization,* a method by which many FM stations broadcast special programming using a 67 kHz subcarrier and reduced amplitude which renders the SCA signal inaudible on normal FM receivers. SCA programming consists of background music, "talking books" for the blind, and special-audience material for physicians and other professionals. The purpose of an SCA decoder is to extract the 67 kHz subcarrier from a received FM signal and demodulate it. Figure 171 shows such a circuit using the 565 PLL IC. The 565 is tuned to 67 kHz by adjusting the 5 K potentiometer. Only rough adjustment of the potentiometer is necessary, since the PLL will "seek" the 67 kHz SCA signal and lock onto it. The input signal should have an amplitude between 80 and 300 mV and be obtained from an input impedance of less than 10 K. The demodulated SCA output is approximately 50 mV and has a maximum frequency response of about 7 kHz. Since much SCA material is either spoken voice or intended for unobtrusive background listening, this is usually not a drawback. An SCA signal is transmitted with about one-tenth the amplitude of an FM stereo signal, and most SCA signals can be received at the distance at which FM stereo reception is possible. While you can listen to the material transmitted via SCA privately, it is illegal to use it "publicly," as for background music at a store, waiting room, or party, without permission of the company or group that produces the SCA material.

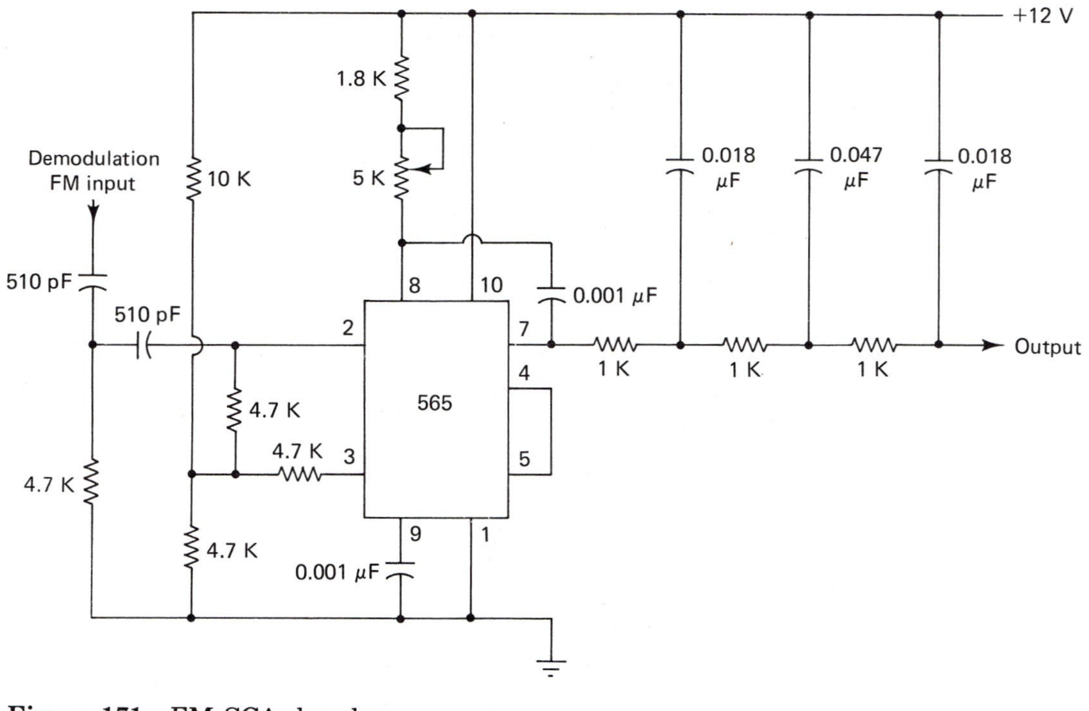

Figure 171 FM SCA decoder

Square root circuit

The circuit in Figure 172 produces an output voltage that is proportional to the square root of an input voltage known as *Vz*. It uses an MC1594 four-quadrant multiplier IC whose inputs are connected together, producing a square root output. *Vz* is a DC voltage between −10 V and 0 V, and the output is given by the formula

$$\text{Output voltage} = 10\sqrt{(Vz)}$$

The circuit is adjusted for proper operation by the two 20 K potentiometers and the 50 K potentiometer. Fixed, known voltage levels are applied as the input *Vz*, and the potentiometers are adjusted for the properly calculated output. (As *Vz* nears 0 V, it may be difficult or impossible to adjust the circuit for proper operation.)

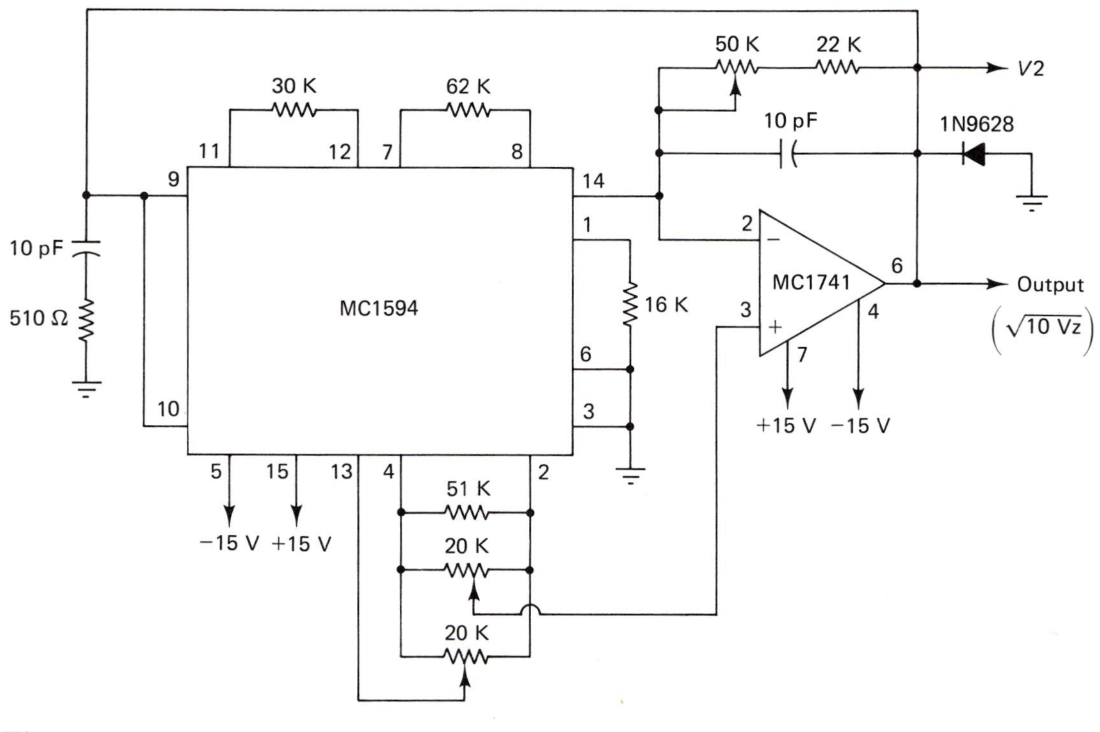

Figure 172 Square root circuit

Squaring circuit

This circuit is essentially a variation of the square root circuit. An MC1594 is configured so that the output is proportional to the square of an input voltage X. The formula for finding the output of the circuit in Figure 173 is

$$\text{Output voltage} = \frac{-X2}{10}$$

where the value of X can range from -10 to $+10$ V. As with the square root circuit, the proper operation of the circuit is determined by adjustment of the potentiometers at different values of X.

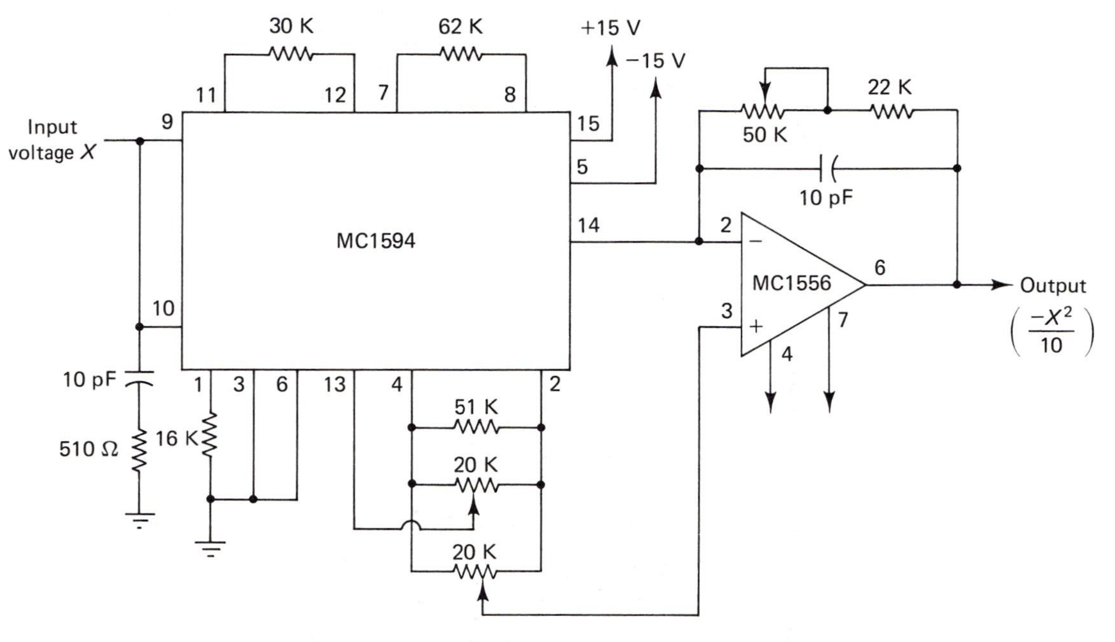

Figure 173 Squaring circuit

Switch, bounceless

Often a constant source of low or high logic levels is needed, but using an ordinary mechanical switch in conjunction with a power source can present problems. When a mechanical switch is closed, a "dirty" logic signal is often produced as the mechanical parts bend and flex when contact is made. This could result in two or more high- or low-level signals being sent to a device's inputs when only one is desired. (In some early microcomputer systems, this caused two letters to be produced when a key was pressed once, and became known as "keyboard bounce.") The solution to the problem is to buffer the mechanical switch with a couple of inverter gates. Figure 174 shows a circuit that uses two sections of a 7404 hex inverter IC which produces a TTL-compatible output, while Figure 175 shows a similar circuit using a 4049 hex inverter for CMOS-compatible output. The switch used can be a pushbutton type if desired.

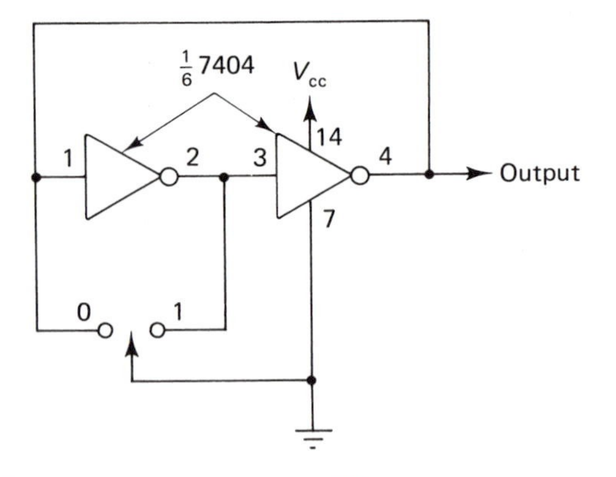

Figure 174 Bounceless switch for TTL

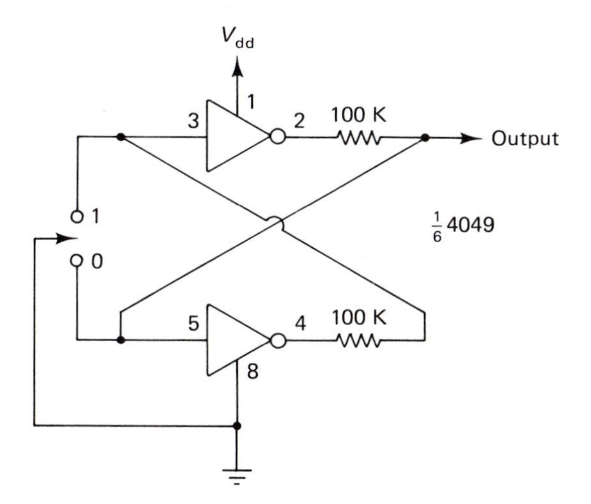

Figure 175 Bounceless switch for CMOS

t

Timebase generator, 60 Hz

Sixty Hertz has become a de facto standard reference frequency in North America, since that is the frequency of AC power line systems and thus is readily obtained. Many devices, such as electric clocks, timers, controllers, etc., are designed to make use of a 60 Hz reference signal in their operation. Thus, there is often a need for a source of this signal independent of the AC power line for portable or battery operation of various circuits, as well as for testing and development. Figure 176 shows a circuit built around the MM5369 timebase generator IC. This device uses a crystal, labeled X in the schematic, cut for the standard 3.579545 MHz color television "burst" frequency, and the outputs of the circuit are both 60 Hz and 3.58 MHz. The 5–50 pF variable capacitor is used to tune the circuit until the outputs are on the correct frequencies as measured by a frequency counter or similar instrument.

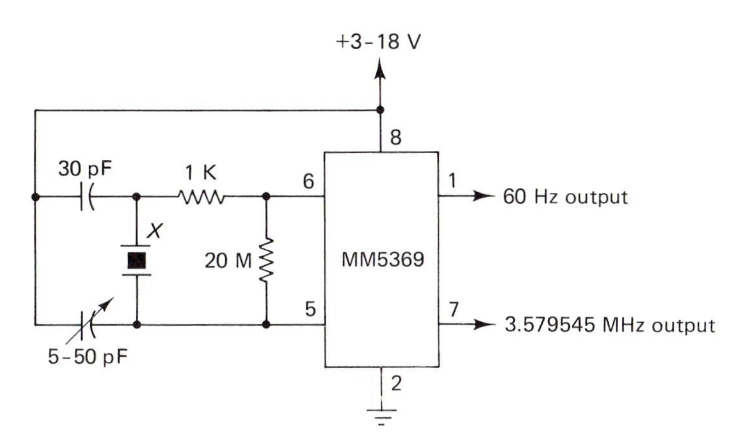

Figure 176 Timebase generator

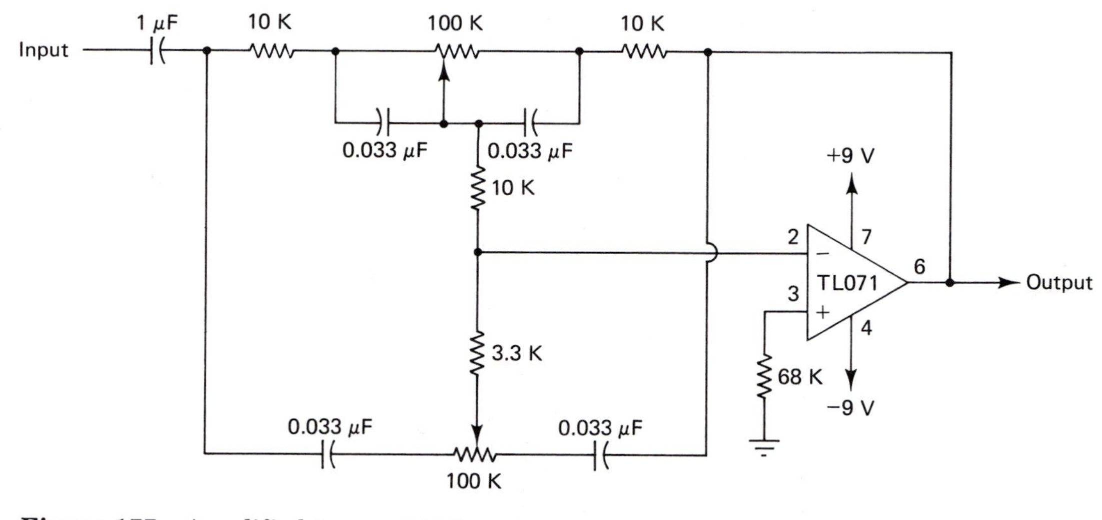

Figure 177 Amplified tone control

Transmitter, FM broadcast band

FCC regulations permit the use of low-powered transmitters in the FM broadcasting band (88 to 108 MHz) without an operator or station license. The circuit in Figure 178 produces a signal near 100 MHz which is audible up to several hundred yards away. The antenna used is an 18″ "whip" type often found on portable radios, while inductor *L* is formed from four turns of #20 wire wound $\frac{1}{4}$ of an inch in diameter and to a length of 5 mm. The output frequency may be altered by changing the dimensions of *L* and the value of the 15 pF capacitor connected across it. The circuit is capable of

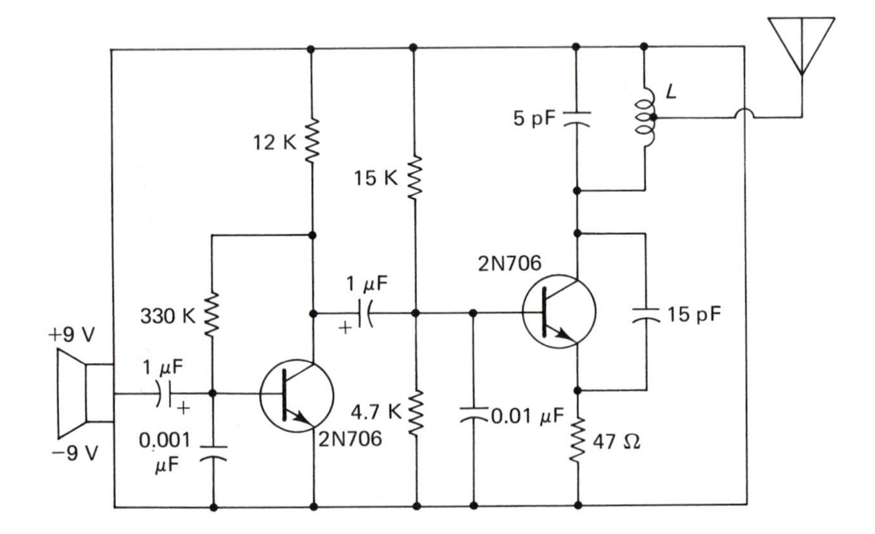

Figure 178 FM transmitter

operating at frequencies up to 150 MHz, depending upon L and the value of the capacitor. (Use on frequencies past 108 MHz requires the appropriate type of license.) The microphone used in this design is an amplified type capable of supplying the dual-polarity 9 V voltages needed by the circuit; a standard type may be used if a 9 V battery powers the circuit.

Transmitter, infrared

This is a companion circuit to the infrared receiver circuit described earlier. The heart of the infrared transmitter, as shown in Figure 179, is an MC14497 PCM remote control transmitter IC driving three MLED71 infrared LEDs. The control input signals may be supplied by individual switches or through the use of a keypad; the eight control input lines permit up to 64 control codes to be transmitted. The scanner output (pin 7) is connected with the row input (pin 3); with some keypads, this is used to release the control inputs into the MC14497. A 500 kHz ceramic resonator is placed between pins 12 and 13 to control the device's internal reference oscillator. The outputs drive the infrared LEDs using a PCM coding scheme which is recognized by the matching infrared receiver. While this circuit was originally designed as part of a TV remote control system, it may be adapted for other uses as well.

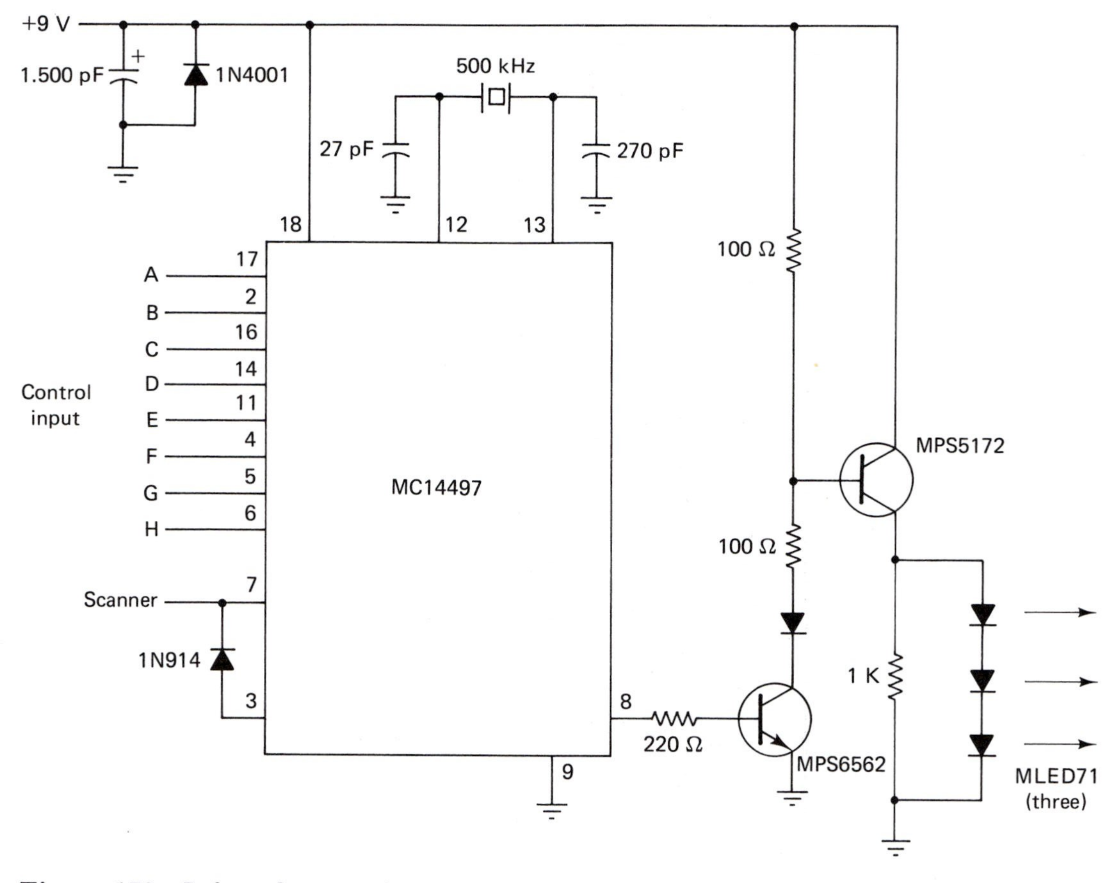

Figure 179 Infrared transmitter

Transmitter, LED

The simple circuit in Figure 180 will take a low-power audio signal, such as that from a transistor radio or tape player, and use it to drive (or "modulate") an LED in accordance with the input signal. The output of this circuit can be received by silicon or selenium solar cells connected to the input of an audio amplifier. The transformer used should have an eight Ω primary and a 500 Ω secondary; the 270 Ω resistor must have at least a 0.5 W rating rather than the common 0.25 W. The diode can be any general-purpose type.

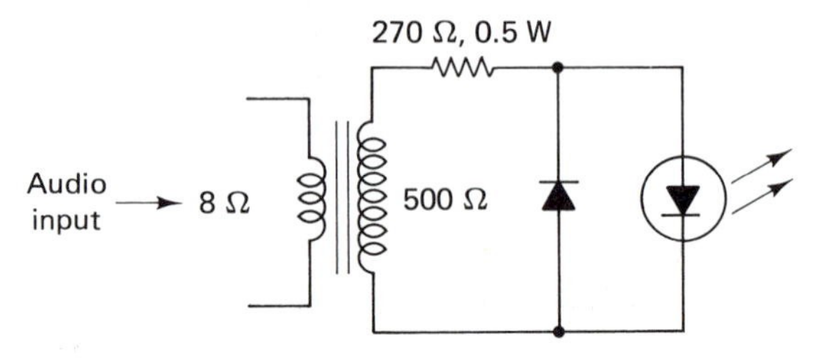

Figure 180 LED transmitter

Transmitter, ultrasonic

The circuit in Figure 181 is a companion to the ultrasonic receiver. Its heart is an MC14457 remote control transmitter device designed for TV remote control. This device has two types of inputs: five row inputs (identified by the prefix R) and four column inputs (identified by the prefix C). These inputs are designed for use with a typical keypad switching matrix; mechanical switches could be substituted

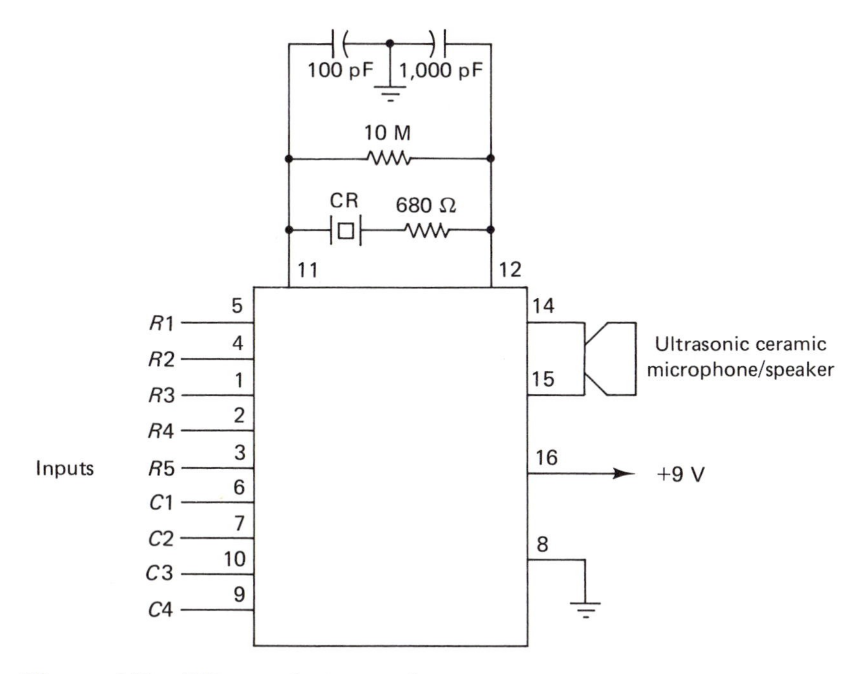

Figure 181 Ultrasonic transmitter

for them if desired. Input data is converted by the MC14457 into frequency-modulated biphase data and is output by the internal oscillator into an ultrasonic microphone or small speaker. The internal reference clock of the MC14457 is controlled by a 500 kHz ceramic resonator (labeled CR in the diagram).

Transmitter, 1760 kHz cordless telephone

Many common cordless telephones use different frequencies for transmission and reception so as to allow "full duplex" operation (both parties can hear and speak at the same time). One common arrangement on older cordless telephones involves using one frequency in the 1,750–1,800 kHz range and another frequency near 49 MHz. The base station unit transmits in the 1,750–1,800 kHz range and receives on 49 MHz; the remote unit does the opposite. The circuit in Figure 182 is a base station transmitter operating on

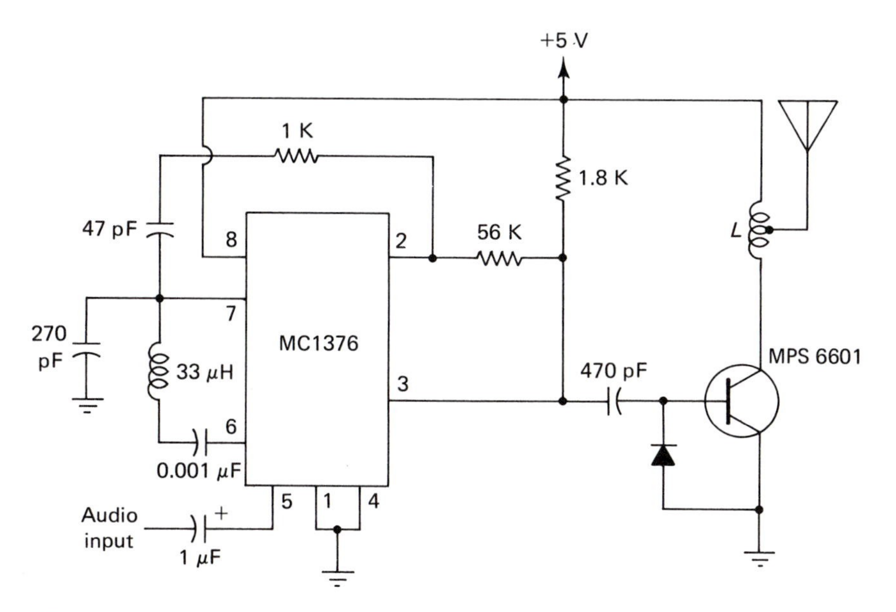

Figure 182 1,760 kHz transmitter

1,760 kHz. The output of this circuit is frequency modulated; the MC1376 IC performs this task. The antenna is a telescoping "whip" type and inductor L can be a "loopstick" such as that often found in AM portable radios as an antenna. (Better results can be obtained if a replacement antenna system for a cordless telephone is used.) The output of this circuit is a few hundred milliwatts.

V

Voltage boosters

Often, "voltage boosting" is done with transformers through the differences in the turns on the primary and secondary windings. However, the output of a transformer is an AC voltage and must be rectified into DC. It is possible to directly convert an AC voltage into a DC voltage that is equal to two, three, or four times the rated RMS value of the AC input. Figure 183 shows circuits that do this boosting. In each one, all capacitors and rectifiers must be rated to *at least twice* the input voltage; for safety, these components should be rated at a good working margin above this minimal level. Large values should be used for the electrolytic capacitors to smooth out the "ripple" present in the output. Polarity of the electrolytics must be observed to prevent damage or "exploding" electrolytics. These circuits must be built and used with care, since, although they are comparatively simple, they can produce very high voltages.

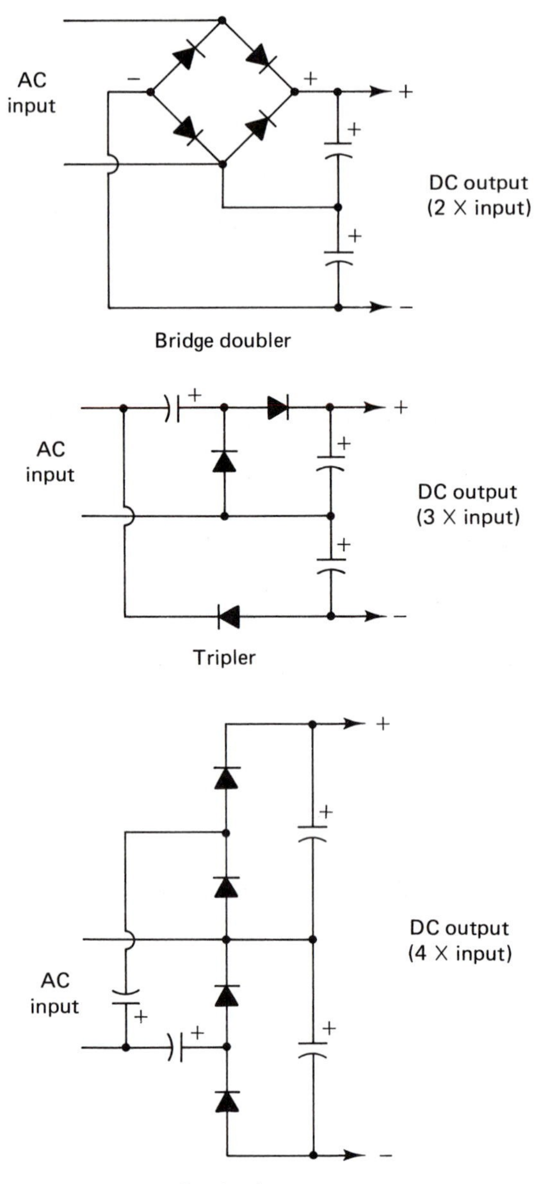

Figure 183 Voltage boosters

Voltage charge pump

Figure 184 shows a novel approach to producing a +12 V output from a +5 V input. The circuit is actually an oscillator built from a 7406 hex inverter IC. The output, when at a high level, is at the standard +5 V TTL level. The 100 μH inductor and 1 μF capacitor serve as "storage tanks" for the electric charge during the high-level output, and discharge during a low-level output. The result is a +12 V output at about 22 mA.

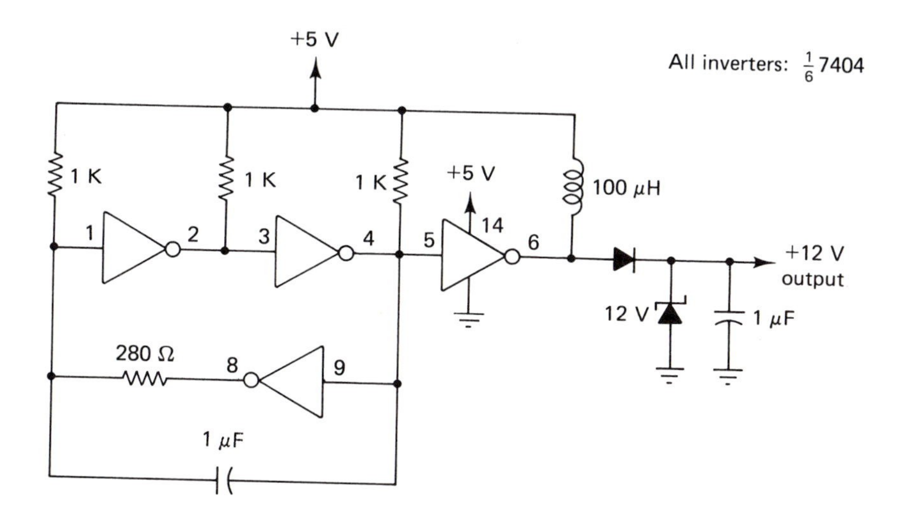

Figure 184 Voltage charge pump

Voltage polarity indicator

A simple method of determining the polarity of an unknown DC voltage is shown in Figure 185. The crucial component in this circuit is the input resistor, which prevents damage to the two LEDs. The proper value of this resistor is found by the formula

$$\text{Input resistor} = \frac{\text{input voltage} - \text{LED voltage}}{\text{LED current}}$$

in which the input voltage used in the calculation should be the largest input voltage the circuit will be used with and the LED voltages and currents are those the LEDs are rated at. If there is any doubt, the input resistor used should be of a larger value than calculated and the circuit not used if the input voltage is not known with reasonable certainty.

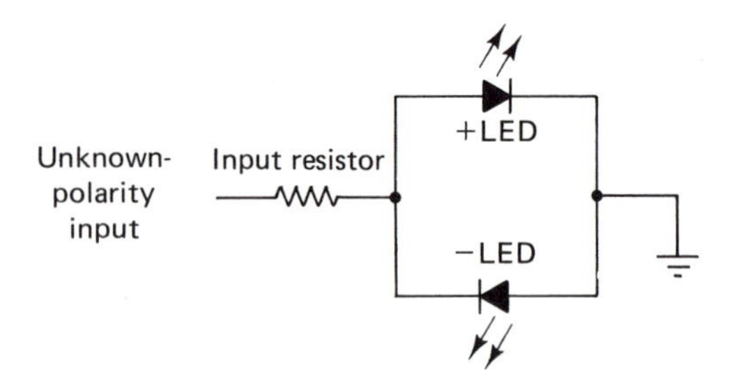

Figure 185 Voltage polarity indicator

Voltage reference

A voltage reference is a voltage source which is constant and stable to a very high degree. Figure 186 shows a simple voltage reference that uses one section of the MC3403 quad operational amplifier, which can operate from a supply voltage of from +3 to 36 V. Using the values of $R1$ and $R2$ shown, the output of the circuit will be equal to half the supply voltage. Thus, a +9 V supply will produce a 4.5 V output. It is possible to alter the ratio of the output to the supply voltage by using different values for $R1$ and $R2$ according to the formula

$$\text{Ratio} = \frac{R1}{R1 + R2}$$

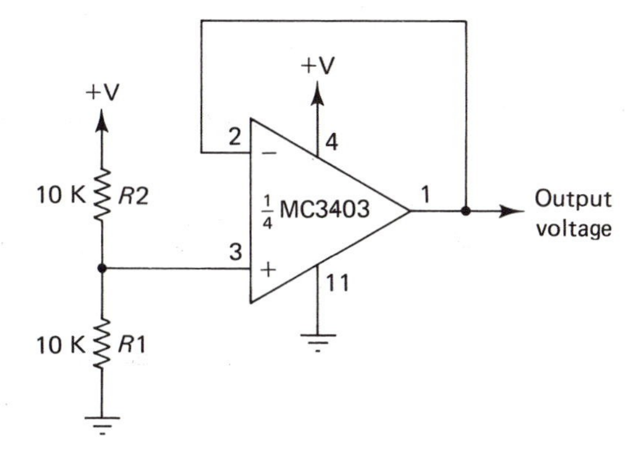

Figure 186 Voltage reference

Voltage regulators

A voltage regulator is a circuit which holds constant an output voltage from a power source over a wide range of current demands by the load connected to the power source. Most voltage regulation today is performed by ICs specifically intended for that purpose which take an input voltage *above* that demanded at output to produce the steady output voltage.

Perhaps the most widely used group of voltage regulator ICs is the so-called 78xx series, made up of the 7805, 7812, and 7815 devices. The last two digits indicate the output voltage supplied by each device (5, 12, and 15 V, respectively). All three devices can accept an input voltage from +2 V greater than the desired output voltage up to +35 V. Each device is also capable of supplying up to 1 A of output current. If the load should unexpectedly demand more current (as would be the case with a short circuit), the 78xx series incorporates on-chip protection circuitry which will temporarily "shut down" the device to prevent its destruction. Figure 187 shows a circuit which can be used with any of the three devices to provide the desired output voltage.

The 78xx series can be used in situations where up to 4 A of

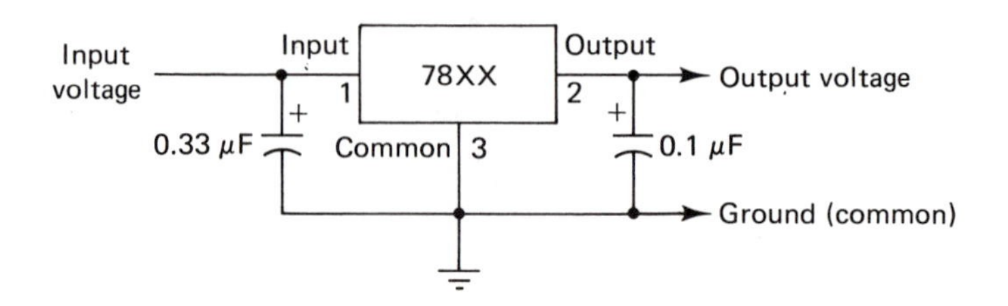

Figure 187 78xx series regulator configuration

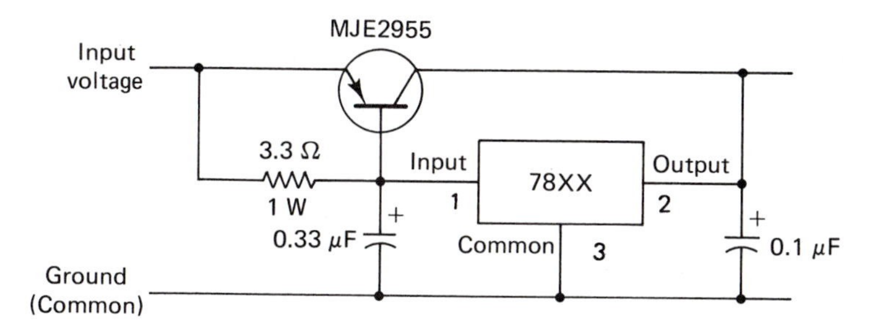

Figure 188 Boosting output current of a regulator

output current is needed if a *pass transistor* is added to the circuit. Figure 188 shows an MJE2955 PNP power transistor added to supply the additional output current. The MJE2955 must be mounted on a heat sink, since the extra current it supplies will cause it to rapidly heat to the point of self-destruction unless some means is provided to cool it down.

It is also possible to "stretch" the output voltage of a 78xx series regulator by using the circuit in Figure 189. The output voltage obtained depends upon the values of $R1$ and $R2$, which are found according to the formulas

$$R1 = \frac{\text{normal output voltage}}{0.02}$$

and

$$R2 = \frac{\text{output voltage needed above normal}}{0.0025}$$

If a variable output voltage is needed, a potentiometer can be used for $R2$.

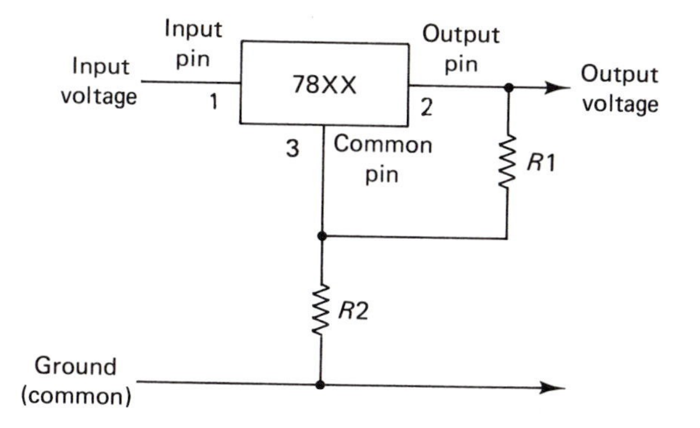

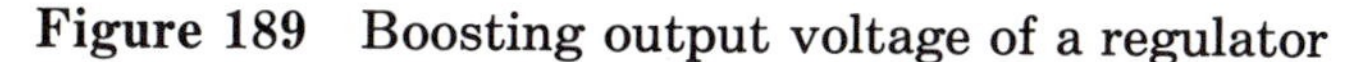
Figure 189 Boosting output voltage of a regulator

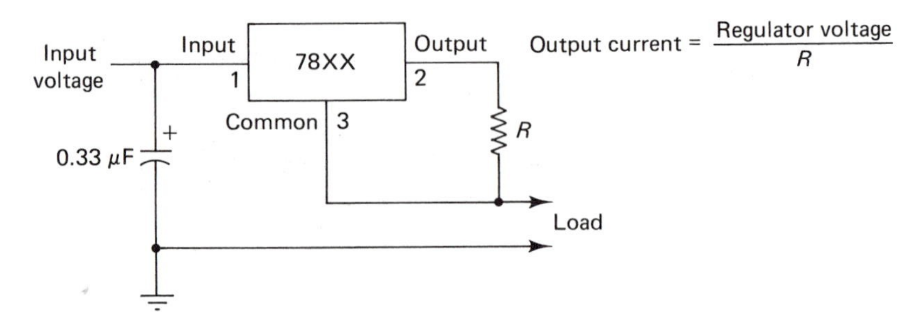

Figure 190 Current regulation

Although intended for voltage regulation, the 78xx series can also be configured to provide a constant level of current. Figure 190 shows how this is done. The output current provided depends upon the value of resistor R, as determined by the formula

$$\text{Output current} = \frac{\text{regulator current}}{R}$$

One problem with power supplies that provide high current is the large amount of heat developed by the pass transistor. This can necessitate large heat sinks and can be a problem if adequate ventilation is not supplied. The solution to this problem is the *switching regulator.* In such a circuit, the pass transistor is rapidly switched on and off (typically, at 5 to 50 kHz); this interrupted duty cycle reduces the heat generated. The result is a square wave output

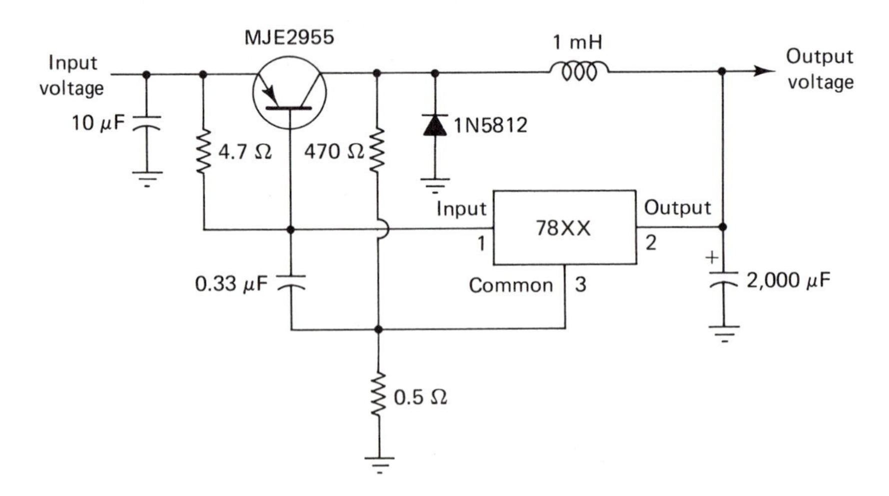

Figure 191 Switching regulator

which is "smoothed" by an inductor. Moreover, since the switching is performed rapidly, the resulting output is easy to filter. Figure 191 shows a switching regulator for the 78xx series which delivers up to approximately 4 A into a connected load. This circuit runs considerably "cooler" than the circuit in Figure 188.

While the 78xx series is very popular, there are additional voltage regulator ICs better suited for other purposes. One is the LM150, which is capable of providing an output voltage over the range of 1.2 to 33 V at up to 3 A of current. The circuit shown in Figure 192 accepts a +28 V input and produces an output of 1.2 to 25 V depending upon the setting of the 5 K potentiometer. A heat sink should be used with the LM150 if much current will be drawn by the regulator's load.

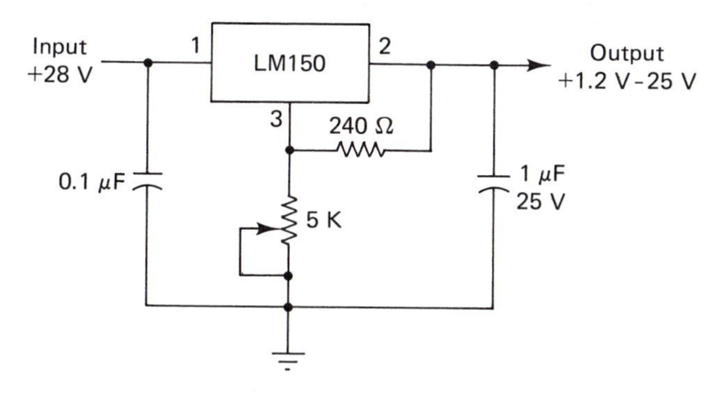

Figure 192 +1.2 V–25 V adjustable output regulator

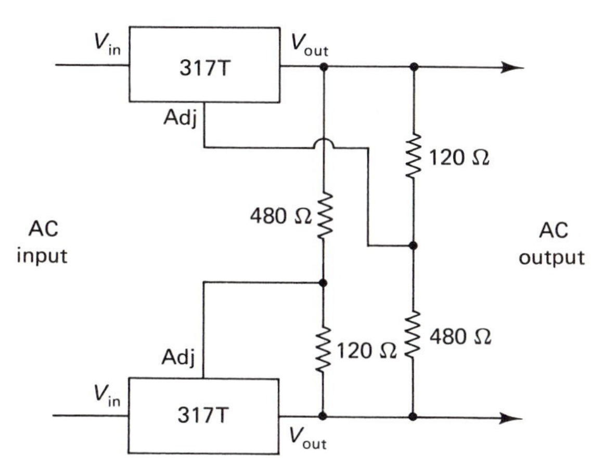

Figure 193 AC voltage regulator

All of the voltage regulators thus far discussed have been designed to be operated exclusively with DC voltages. However, there may be times when a similar action is desired for AC voltages. One approach is illustrated in Figure 193, which shows an AC voltage regulator using two 317T devices. A single 317T is capable of supplying an output voltage over a range of +1.2 to 37 V at up to 1.5 A. By combining two such devices so that one can handle one-half of an AC input signal, the circuit shown can provide a 6 V peak-to-peak signal at up to 1 A of current from a 12 V peak-to-peak input waveform. However, the output of this circuit will not be a pure sine waveform; rather, it will more closely resemble a square wave with sloping sides. As such, the circuit will introduce considerable distortion and is not suitable for application where a well-formed sine wave signal is important.